ECONOMIC COMMISSION FOR EUROPE
Geneva

Emerging Global Energy Security Risks

ECE ENERGY SERIES No. 36

UNITED NATIONS
New York and Geneva, 2007

NOTE

The designations employed and the presentation of the material in this publication do not imply the expression of any opinion whatsoever on the part of the Secretariat of the United Nations concerning the legal status of any country, territory, city or area, or of its authorities, or concerning the delimitation of its frontiers or boundaries.

Mention of any firm, licensed process or commercial products does not imply endorsement by the United Nations.

UNITED NATIONS PUBLICATION

Sales No. E.07.II.E.22

ISBN 978-92-1-116975-1
ISSN 1014-7225

FOREWORD

The security of energy supplies has been a priority for the UNECE on a number of occasions since the Commission was founded in 1947. After the Second World War, UNECE established an allocation system to alleviate post-war coal shortages. After the 'energy crisis' in the 1970s, east-west energy trade and cooperation became an important alternative for UNECE member States to oil and natural gas imports from the Middle East. Today, energy security has once again risen to the top of the economic and political agenda because of volatile oil prices, sharply rising demand for oil and gas, international tensions and terrorism. Importing countries are again seeking greater security through the diversification of energy supply sources, including from the Caspian Sea region.

In response to these developments, the UNECE Committee on Sustainable Energy launched the Energy Security Forum (ESF) in 2003 to assess risks and to develop and promote risk mitigation options together with senior representatives of the energy supply industries, financial institutions and relevant international organizations. Since then the Forum has appraised a wide range of new energy security threats and reviewed the impact these could have on energy and financial markets.

During the 2005 session of the Committee on Sustainable Energy, the ESF was requested to provide its views on global energy security risks to the host authorities of the G8 Summit Meeting held in St. Petersburg, July 2006. At the same time, high-level government representatives of the Republic of Azerbaijan, the Republic of Kazakhstan, the Islamic Republic of Iran, the Russian Federation and the Republic of Turkey agreed to work with the Forum to examine the potential contribution of increased energy exports from the Caspian Sea region to provide greater diversity of energy supply sources to UNECE member States.

This publication is the result of the analyses and discussions conducted by the Energy Security Forum on the global dimensions of emerging energy security risks facing UNECE member States. It presents global energy security risks from three different points of view: the European Union, the Russian Federation and from North America. These views were reconciled during the deliberations of the Energy Security Forum that arrived at a consensus on the conclusions and recommendations on how to mitigate these risks.

This publication also examines how the Caspian Sea region can contribute to energy supply diversification. It reviews the energy transport corridors, new infrastructure, transmissions systems and investment requirements needed to accomplish this.

Addressing the challenges identified in this publication will require a renewed commitment from UNECE countries for a broadly based intergovernmental dialogue on the principles underlying energy relationships and trade, on energy security and on sustainable energy development. The Committee on Sustainable Energy, with guidance from the Commission, has decided to undertake this new dialogue.

During the last sixty years, the UNECE has offered a forum to governments of the region with different and sometimes conflicting interests, to meet and develop common understandings and move forward on international issues of mutual interest. In the years ahead, the ECE will continue to offer this neutral platform to resolve new priorities as they emerge.

It is my pleasure to bring to your attention this UNECE publication on Emerging Global Energy Security Risks and to pay tribute to Dr. George Kowalski, Director of the UNECE Sustainable Energy Division for his significant contribution to this effort.

Marek Belka
Executive Secretary
United Nations Economic Commission for Europe

Emerging Global Energy Security Risks

Table of Contents

List of Tables

List of Figures

List of Boxes

List of Boxes

PREFACE

The current concern over energy security has been at the forefront of the preoccupation of UNECE member States since at least 2000. Over the last seven years, various factors have heightened concerns and added to anxieties regarding energy availability and security of energy supplies, including: rapid economic growth; increasing dependence on external energy supplies; Middle East political tensions; sabotage and terrorist attacks; the 2003 electric power blackouts in North America and Europe; the interruption of natural gas and oil supplies in early 2006 and early 2007 respectively in Europe; the forced re-negotiation of oil revenue sharing arrangements between governments and the private sector in some oil producing countries; and conflicts in a number of crude oil and natural gas producing regions.

Despite the growing public concern and efforts by countries to develop a common understanding of energy security risks and risk mitigation strategies, there continue to be wide differences among UNECE member States on key aspects of energy security, including their causes and appropriate policy responses. The inability of countries to forge a common approach on energy security is due to the significant divergence in the energy mix, industry structure, and availability of domestic energy resources, particularly of crude oil and natural gas, among countries; differences in access to alternative energy imports, geopolitical influence and energy policy orientations; and the differences in capacity, disposition and willingness of countries to deal with international issues on a bilateral and multilateral basis.

With a view to enhancing dialogue and promoting the possible convergence of views on energy security issues in the UNECE region, the Committee on Sustainable Energy has addressed the subject on a periodic basis through high-level meetings, the publication of CD-ROMs and the establishment in 2003 of the UNECE Energy Security Forum to better engage the private sector, both the energy industries and the financial community, on this matter.

As part of the ESF initiative, three research papers from leading experts on energy security were commissioned and a number of workshops organized to discuss the conclusions and recommendations of these three papers. Likewise, a report on the contribution of the Caspian Sea region to mitigating global energy security risks was prepared, benefiting from contributions from the countries of the Caspian Sea region, a high-level meeting and a seminar on the topic.

These reports and the proceedings of the ESF meetings comprise the contents of this publication: Emerging Energy Security Risks and Risk Mitigation in a Global Context (Section 1) and the Report on Global Energy Security and the Caspian Sea Region (Section 2). A number of the findings from these reports were transmitted by the private sector members of the ESF to the Government of the Russian Federation as host of the G8 Summit in St. Petersburg. This was carried out in response to the request of the Special Envoy of President Vladimir Putin on International Energy Cooperation that was made at the annual meeting of the Energy Security Forum, June 2005, held in conjunction with the annual session of the UNECE Committee on Sustainable Energy.

.../

This publication also presents conclusions and recommendations from these various contributions, as well as the insights of ESF members on global energy security risks and risk mitigation. It should be noted that the focus of this volume is solely on supply-side issues related to energy security. Demand management and energy efficiency issues are not explicitly addressed. Clearly, the best energy is the energy that is saved and not consumed. Therefore, one of the best ways to enhance energy security is to improve energy conservation and the efficiency of production, transport and use of energy. Since the potential in the UNECE region and most notably in Eastern Europe and Central Asia is large, policy initiatives to improve energy security should start with measures to improve energy conservation and energy efficiency.

For further information about the Energy Security Forum, please contact:

United Nations Economic Commission for Europe
Palais des Nations, CH-1211 Geneva 10, Switzerland
Telephone: (41-22) 917-2451 or 917-1234
Telefax: (41-22) 917-0038 or 917-0123
alexandre.chachine@unece.org
http://www.unece.org/ie

ACKNOWLEDGEMENTS

This publication has been prepared at the request of the UNECE Committee on Sustainable Energy. It is drawn from the work of the Energy Security Forum since its formation in 2003 and from its Executive Committee meetings, annual sessions, seminars and workshops. The study on emerging energy security risks and risk mitigation in a global context in Part 1 of this book and the report on global energy security in the Caspian Sea region in Part 2 were requested by government representatives during a High-Level Meeting held during the 2005 annual session of the Energy Security Forum (see Annex) and mandated by the Committee on Sustainable Energy (ECE/ENERGY/65).

The secretariat would like to acknowledge the contribution of elected officers, members and donors of the Energy Security Forum in the preparation of this report. In particular, the secretariat would like to acknowledge the special contribution of Mr. Robert McFarlane, Chief Executive Officer, Energy and Communication Solutions, and Mr. Togrul Bagirov, Executive Vice President, Moscow International Petroleum Club. Their leadership and commitment have been invaluable in the work and accomplishments of the Energy Security Forum. The secretariat would also like to acknowledge the important contribution of three reports commissioned for this publication, which were prepared by Mr. Chris Lambert, Director, Westminster Energy Forum, London (Section 1.4); Dr. Yury A. Ushanov, under the scientific guidance of Mr. Andrei A. Kokoshin, Academician, Russian Academy of Sciences, Moscow (Section 1.5); and Mr. Terry Newendorp, Chief Executive Officer, Taylor-deJongh, Washington D.C. (Section 1.6).

In addition, the secretariat gratefully recognizes the participation in Energy Security Forum annual sessions, seminars and workshops since 2003 of senior representatives of the American Petroleum Institute (API), Aon Energy Group, Aon Risk Consulting, Azpetrol, ChevronTexaco, ConocoPhilips, Core Ratings, Credit Suisse First Boston, Energy and Communications Solutions LLC, Energy Charter Secretariat, European Commission, ExxonMobil, Fitch Ratings, Framatom ANP, OAO Gazprom, International Energy Agency (IEA/OECD), International Energy Forum (IEF), ITERA Oil and Gas Company, KazMunayGaz, Lehman Brothers International, Litasco, Lukoil, Moscow International Petroleum Club, Organization of the Petroleum Exporting Countries (OPEC), Organization for Security and Cooperation in Europe (OSCE), RAND Corp., Ruhrgas AG, Shell International Ltd., StroyTransGas, Swiss Re, Total, Transneft and the Westminster Energy Forum. The contributions of all the aforementioned are much appreciated by the Committee on Sustainable Energy and the UNECE secretariat.

ACKNOWLEDGEMENTS

[illegible]

[illegible]

[illegible]

Emerging Energy Security Risks and Risk Mitigation in a Global Context

1.1 EXECUTIVE SUMMARY

Global energy security risks have increased sharply because of steeply rising oil import demand in developed and more importantly developing countries; the narrowing margin between oil supply and demand which has driven up prices; the volatility of oil prices arising from international tensions, terrorism and the potential for supply disruptions; the concentration of known hydrocarbon reserves and resources in a limited number of the world's sub-regions; the restricted access to oil and gas companies for developing hydrocarbon reserves in some countries; the rising cost of developing incremental sources of energy supplies; the lengthening supply routes; and the lack of adequate investment along the energy supply chain, including the electricity sector.

Governments in producing and consuming countries can mitigate these risks by promoting investment in the energy sector through the provision of the legal frameworks, regulatory environments, tax incentives together with fair and transparent processes to foster the public-private partnerships needed to promote and protect investments in existing and new oil and natural gas supplies; by removing barriers to trade and investment for both private sector and public energy companies; by encouraging the mutual self-interest of energy producers and consumers to secure long-term and committed demand for hydrocarbons; and by seeking the convergence of norms, standards and practices as well as new forms of cooperation to facilitate the financing of resource developments.

Additionally, government measures are needed to promote energy security that complement, flank and facilitate the functioning of markets. Energy security risks and rising import dependence can be mitigated by a range of additional policy options aimed at furthering the diversification and flexibility of energy systems, including multiple supply routes; increasing indigenous (domestic) energy supplies; improving energy conservation and efficiency; expanding the fuel mix available to consumers; diversifying energy sources of supply; building-up and maintaining strategic and commercial stocks where warranted; encouraging research and development in greening the fossil fuel energy supply chain; developing and commercializing new and renewable sources of energy; improving the protection and safety of energy infrastructure against possible acts of terrorism; and strengthening international cooperation.

The strengthening of policy measures and the mitigation of energy security risks would benefit to a significant degree from a strengthened and more coordinated multilateral producer-consumer dialogue between governments, industry, the financial community and relevant international organizations on the following issues: (a) data and information sharing and increased transparency, (b) infrastructure investment and financing, (c) legal, regulatory and policy framework, (d) harmonisation of standards and practices, (e) research, development and deployment of new technologies, and (f) investment/transit safeguards and burden sharing.

There is already considerable work underway in many of the areas identified above, not only at the UNECE but also in other international organizations, such as the International Energy Agency (IEA/OECD), the International Energy Forum (IEF), the Energy Charter and the Organization of the Petroleum Exporting Countries (OPEC). Nonetheless, these ongoing activities could benefit from stronger multilateral cooperation and political endorsement.

Many of the elements of the UNECE programme of work in energy are of direct or indirect relevance to the issues raised above. In addition, the UNECE Committee on Energy and some of its subsidiary bodies have directly addressed energy security issues periodically over many years. The Committee continues to be well placed for a pan-UNECE dialogue on energy security issues and related aspects, including the relationship between financial markets and energy security.

1.2 CONCLUSIONS AND RECOMMENDATIONS

The Energy Security Forum, which comprises members of the energy industries and financial sector working under the auspices of the UNECE, undertook numerous analyses and discussions on the global dimensions of emerging energy security risks facing UNECE member States. In particular, energy security risks were assessed from three different points of view: the European Union, the Russian Federation and from North America. These views were then reconciled during deliberations of the ESF that arrived at a consensus on the conclusions and recommendations on how to mitigate these risks. These agreed Conclusions and Recommendations are provided in Box 1.2.1.

Furthermore, during the 14th Annual Session of the UNECE Committee on Sustainable Energy held in Geneva in June 2005, the ESF was asked to deliver its views on global energy security risks to the host authorities of the Group of Eight (G8) Summit Meeting held in St. Petersburg in July 2006. In response to this request by the Russian Federation, the ESF, under the auspices of the UNECE, submitted the set of Conclusions and Recommendations outlined in Box 1.2.1.

At the 15th annual session of the Committee on Sustainable Energy in November 2006, delegations recommended that the Committee continue its work in this area and undertake a broadly shared intergovernmental expert dialogue on energy security issues identified by the ESF and outlined in Box 1.2.1.

Box 1.2.1 Conclusions and Recommendations

CONCLUSIONS AND RECOMMENDATIONS

BASED ON THE RESULTS OF THE ESF STUDY ON EMERGING ENERGY SECURITY RISKS AND RISK MITIGATION IN A GLOBAL CONTEXT

Emerging global energy security risks stem from a complex diversity of political, social, economic, financial, legal, geographic and technical factors, including ongoing civil strife, ethnic conflicts and growing international tensions; these also include international terrorism as an important factor menacing global energy security;

The faster than expected increase in oil imports among developing countries; the potential for supply disruptions; the concentration of known hydrocarbon reserves and resources in a limited number of the world's sub-regions; restricted access to oil and gas companies for the development of hydrocarbon reserves in some countries; and the perceived long run scarcity of hydrocarbon resources are among key-factors contributing to global energy security risks;

continued ... /

Box 1.2.1 Conclusions and Recommendations *continued*

The mitigation of these risks would benefit to a significant degree from a multilateral producer-consumer dialogue between governments, industry, the financial community and relevant international organisations based on the reciprocal interests of producers and consumers within the framework of the G8 countries and the United Nations, particularly under the auspices of the UN Economic Commission for Europe which has had an energy programme since 1947;

An advantage of the United Nations lies in its capacity to address the diversity of emerging energy security risks with all parties concerned and in its ability to convey recommendations, when appropriate, to the attention of the governments of member States, the Economic and Social Council and the Security Council;

In contrast to the long lead times associated with the development of new energy supply solutions, such a dialogue can provide risk mitigation options in the short-term, reinforcing the interdependence between producers and consumers alike, but this needs to be a continuous exchange rather than periodic in order to ensure the depth of discourse and consistency of advice;

Acute global risks are emerging from the lengthening supply routes that are becoming distinctly more vulnerable to sabotage and terrorist attack; rising cost of developing incremental sources of energy supplies; infrastructure constraints along the energy supply chain, including the electricity sector, due to lack of adequate investment; and the unresolved social and ethnic disputes in a number of producing and transit countries;

These energy security risks and rising import dependence can be mitigated by a range of policy options aimed at furthering the diversification and flexibility of energy systems, including multiple supply routes; increasing indigenous energy production; improving energy conservation and efficiency; expanding the fuel mix available to consumers; diversifying energy sources of supply and transportation routes; building-up and maintaining strategic and commercial stocks where warranted; encouraging research and development in greening the fossil fuel energy supply chain; developing and commercializing new and renewable sources of energy; and strengthening international cooperation;

Within the current range of energy prices and with the present technology, reserves of crude oil and natural gas are expected to be capable of meeting cumulative world demand over the next forty or more years;

Although hydrocarbon fuel reserves are largely sufficient to meet energy demand for many decades to come, global energy security risks arise from their geographic distribution and high concentration, frequently in economically vulnerable and unstable regions of the world, providing no guarantee that those hydrocarbons will be accessible and delivered when needed;

Ensuring the security of hydrocarbons supply requires access to and development of reserves; availability, access to and reliability of transportation and all related infrastructure; appropriate legal, regulatory, fiscal and policy frameworks that are conducive to investment in both producing and consuming countries; technology transfer to enhance the efficiency and recovery of energy resources and acceptable methods of addressing environmental issues;

Promoting the mutual self-interest of energy producers and consumers to secure the long-term and committed demand for hydrocarbons while seeking convergence of norms, standards and practices as well as new forms of cooperation can help to facilitate the financing of resource developments and thereby enhance the security of energy supply;

Emerging global energy security risks could be greatly diminished by concerted international efforts to remove barriers to trade and investment for both private sector and public energy companies so that they can effectively and efficiently use capital to develop energy reserves and resources;

Governments in both consumer and producer countries need to provide a legal framework, regulatory environment, and tax system together with fair and transparent processes to foster the public-private partnerships needed to promote and protect investments in new oil and natural gas supplies and in enhanced secondary and tertiary recovery of hydrocarbons;

continued ... /

Box 1.2.1 Conclusions and Recommendations *continued*

The crucial role of petroleum as a transportation fuel is another emerging energy security risk that can be addressed with demonstrably effective measures including inter-fuel substitution to bio-fuels and natural gas, in particular LNG (liquefied natural gas), the use of new lightweight materials in vehicle construction and further development of hybrid vehicles;

Each policy for enhancing energy security has a corollary in the policy measures for promoting sustainable energy development in both importing/consuming and exporting/producing countries;

A multilateral dialogue on the security of energy supplies and the security of energy demand needs to be structured with an international energy risk matrix to identify, categorise and develop strategies to mitigate energy security risks; appraise rapidly rising global energy demand; and reduce barriers to domestic and foreign investments in key infrastructure, oriented towards sustainable energy development, within the framework of the United Nations is essential for ensuring secure, reliable and affordable energy resources while reducing environmental pollution;

While acknowledging that some of the foregoing matters are addressed by the UN Economic Commission for Europe, International Energy Agency (IEA/OECD), International Energy Forum (IEF), Energy Charter and the Organization of the Petroleum Exporting Countries (OPEC), the Energy Security Forum recommends that the governments of the G8 commit themselves to a broadly shared multilateral producer-consumer dialogue in the following areas:

(a) data and information sharing and increased transparency;
(b) infrastructure investment and financing;
(c) legal, regulatory and policy framework;
(d) harmonisation of standards and practices;
(e) research, development and deployment of new technologies; and
(f) investment/transit safeguards and burden sharing.

1.3 ENERGY SECURITY RISKS AND RISK MITIGATION: A GLOBAL OVERVIEW

Defining the Security of Energy Supply

Although energy security is currently one of the most debated issues in the UNECE region, a generally accepted definition is still lacking. Therefore, the term "energy security" or "security of energy supplies" is used in various contexts, for different purposes, often having very dissimilar meanings. While energy security is not easy to define because it is a multifaceted concept, there are four dimensions of particular relevance: (a) physical disruption of supplies due to infrastructure breakdown, natural disasters, social unrest, political action or acts of terrorism; (b) long-term physical availability of energy supplies to meet growing demand in the future; (c) deleterious effects on economic activity and peoples due to energy shortages, widely fluctuating prices or price shocks; and (d) collateral damage from acts of terrorism resulting in human casualties, serious health consequences or extensive property damage. All four dimensions are relevant in the current environment.

Taking into consideration these four dimensions, energy security could be defined as "the availability of usable energy supplies, at the point of final consumption, at economic price levels and in sufficient quantities and timeliness so that, given due regard to encouraging energy efficiency, the economic and social development of a country is not materially constrained". Clearly, this is but one of a number of possible definitions that could be put forward, however it does have the merit of capturing the multidimensional nature of energy security.

Due to the complexity of the issue and its multidimensional nature, the work programme of the Energy Security Forum focused primarily on one element of energy security, that is, the long- term physical availability of energy supply to meet growing future energy demand. The components examined included the future availability of energy resources, the reliability of energy suppliers, deliverability through infrastructure and affordability at the consumer end. The other dimensions of energy security were not examined and, therefore, are not part of this report. For example, the macroeconomic consequences of energy disruptions or price shocks, the vulnerability of energy infrastructure to terrorism and so on were not assessed and are therefore not discussed in this publication.

The Role of Markets

It is commonly accepted that economic efficiency is best promoted through decentralized and liberalized energy markets, with freely determined market prices. Over the last ten to fifteen years, technological, institutional and societal changes have tended to favour the implementation of measures to open up and liberalize energy markets. However, there continues to be a wide diversity of views among countries on the role of free markets and market forces in promoting societal objectives, such as, energy security.

Due to geopolitical, economic and historic considerations, the belief in free markets and the power of free markets to deliver on social objectives is strongest in North America. The view in Western and Central Europe is more varied. Some countries have a predilection or predisposition to market solutions while others favour a more cautious approach with strong government oversight and intervention whenever needed. For example, despite the efforts and the vigorous measures taken by the European Commission to open up and liberalize electricity and natural gas markets in the European Union (EU) region, a number of EU governments continue to be attached to their national state enterprises, to favour national champions and to closely oversee the functioning of energy markets.

The belief or commitment to free markets is much less pronounced in countries of Eastern Europe and Central Asia for a variety of reasons, though here again there is no unique view. For example, the free interplay of market forces in the Russian Federation is somewhat constrained by government measures favouring the creation of large state owned or controlled enterprises in the oil and gas sector, state control of oil and gas pipeline facilities, particularly export pipelines, and the imposition of limits on the foreign ownership/control of energy assets while, at the same time, accepting some private-sector ownership of energy assets. Kazakhstan, on the other hand, has been more open to the development of energy resources by the private sector. Nonetheless, it is probably fair to conclude that the commitment to free markets in energy is less pronounced in Eastern Europe and Central Asia than in Western Europe and North America.

Consequently, national aspirations and the diversity of views on the role of the market and of government, including the different market practices and institutional arrangements in countries, complicate discussion of and agreement on collective efforts to improve energy security. In addition, the private sector, while recognizing the role of government in establishing investment conditions that are fair and conducive to facilitating inward investment flows, is on the whole much less predisposed to direct intervention in energy markets.

Large oil and natural gas companies, private as well as state-owned, have had a significant influence and played a major role in the development of the world's hydrocarbon industry in the past. However, there are now concerns that in a period of heightened instability and with the rapid growth in energy demand in developing countries, the private sector as well as state companies may not be able, by themselves, to ensure sufficient energy supplies to meet the growing demand for energy in the future and to prevent disproportional energy price increases that could send shock waves through national economies. The sharp increases in crude oil, natural gas and electricity prices worldwide since 2002 are seen as a sign or motivator for more government involvement and intervention in energy markets to ensure access to and the development and deliverability of energy resources, notably hydrocarbons, at economic price levels.

However, not everyone shares this opinion. There are those that believe that the massive investment of about US$ 20 trillion in energy infrastructure worldwide that will be required over the next three decades, according to the International Energy Agency (IEA)[1], can only be raised through the efficient and unhindered functioning of markets. This is predicated on the view that the current growing energy security concerns are at least partially the consequences of long-standing market inefficiencies, the lack of suitable, transparent and favourable investment frameworks and excessive state intervention in many energy producing and transit countries.

Irrespective of one's views on the role of the market and of government, it would seem that a strengthened dialogue on energy security, its principles and policy alternatives, among countries within the UNECE region would be worthwhile. Many UNECE countries are alarmed by the expected increase in their crude oil and natural gas import dependency.

While this increasing and high dependency itself does not necessarily reflect a deterioration in energy security, importing countries nonetheless are uncomfortable with the thought of being reliant on a few suppliers for their energy needs. This is particularly the case for hydrocarbons where the major suppliers are the Russian Federation, the Organization of the Petroleum Exporting Countries (OPEC) and countries from the Caspian Sea region and Africa. Despite the relatively good historical reliability of crude oil and natural gas deliveries from those regions, the lack of substantial domestic crude oil and natural gas reserves, intensified competition with emerging economies, such as the People's Republic of China and the Republic of India, and persistent high energy prices, have created an uneasy feeling about energy security in many UNECE countries.

This greater sense of energy security vulnerability is leading countries to search for ways and means to enhance their security of energy supplies. On the other hand, producing countries, such as the Russian Federation, the Kingdom of Norway, the Caspian Sea producers and others are seeking greater security of demand. Developers are called upon to make large upfront capital commitments in the hope that demand and prices remain reasonable over the life span of projects that are usually in the order of 30 to 40 years. This mutuality of interests suggest that a regional dialogue or compact among UNECE countries on the subject could be meaningful and could lead to policy measures that would benefit both consuming and producing countries.

Hydrocarbon Resources

The re-emergence of concern over high oil and natural gas prices and apprehension over security of oil and natural gas supplies has rekindled the fear that the world could soon run out of natural resources, notably hydrocarbons (oil and natural gas). Once again, stark warnings, similar to those heard in the 1970s, can be heard about the sharp draw down of conventional hydrocarbon resources.

Within the current range of energy prices and with the present technology, it is estimated that conventional reserves of crude oil and natural gas are expected to be capable of meeting cumulative world demand for the next forty or more years. The current worldwide reserve to production ratios of 40 to 70 for crude oil and 70 to 100 for natural gas provide a comforting picture in that respect. In addition, there are large non-conventional hydrocarbon resources that could be developed, if necessary, to meet growing demand, notably for oil. Hence, resource depletion per se is not of major concern at this time. However, what is of vital concern is whether the existing and potential new reserves will be financed and developed in an efficient and timely manner. This is the unanswered question that currently hangs over oil and natural gas markets, adding to uncertainty, risks and anxieties.

While global fossil fuel reserves, including those of hydrocarbons, are sufficient to meet energy demand growth for many decades to come, their unequal geographic distribution and high concentration in several vulnerable and unstable regions of the world is generating concern about whether those hydrocarbons will be accessible, developed and delivered when needed. By 2030, the Middle East is expected to supply around 40 per cent of all the oil consumed in the world, compared to about 30 per cent now. OPEC is likely to supply about 50 per cent by 2030, which is close to the 54 per cent share it supplied in 1973 during the first oil crises, as compared to the today's 40 per cent share.[1] Moreover, approximately two-thirds of the world's established reserves of crude oil are in the Middle East. While gas reserves are not as highly concentrated geographically as oil reserves, nonetheless two countries, the Russian Federation and the Islamic Republic of Iran, have about 40 per cent of the world's reserves.

In addition, with the high geographic concentration of hydrocarbon reserves, direct access by large international oil and gas companies to those reserves and to hydrocarbon resources is increasingly being restricted. Currently, more than 75 per cent of the world's hydrocarbon reserves lie outside their reach. And therefore, lacking investment opportunities in upstream projects, more and more of those companies are redirecting their considerable earnings away from exploration and upstream development to repurchasing their own shares and/or increasing dividends to shareholders.

According to the International Energy Agency, about US$ 8 trillion of investment will be needed globally over the next three decades to maintain and expand energy supply systems in the oil and natural gas sectors and, most notably, in upstream oil and gas projects. The problem though is that most of the remaining undeveloped hydrocarbon reserves and resources are concentrated in developing countries. Unfortunately, many of these countries are not private sector investment friendly and, moreover, as mentioned earlier are in economically vulnerable and unstable regions of the world.

Consequently, it can be concluded that hydrocarbon reserves and resources are sufficient to meet the growing demand for energy over the foreseeable future. Likewise, financing is available. However, the environment for developing these reserves is not sufficiently investment friendly at the current time. Ensuring the security of hydrocarbon supplies will require access to and development of these reserves; availability, access to and reliability of transportation infrastructure; and appropriate legal, regulatory, fiscal and policy frameworks conducive to investment; technology transfer to enhance the efficiency and recovery of energy resources; and acceptable methods of addressing environmental issues. For this to happen, UNECE countries, both individually and collectively, will need to engage hydrocarbon producing countries to tackle domestic problems and remove existing barriers to investment while, at the same time, taking active measures to mitigate against the potential risks of inadequate future hydrocarbon supplies.

While long discussed and feared, the pressure exerted on world energy markets and particularly on the hydrocarbon markets by the emerging fast-growing economies, such as the People's Republic of China, the Republic of India and the Federative Republic of Brazil, has now materialized. Over the last three to four decades, hydrocarbon market disturbances have tended to originate on the supply side, but this time demand pressures have also contributed to the current disturbances and the tight market conditions. Furthermore, the increased demand for crude oil and natural gas by emerging economies has intensified the direct competition with UNECE countries for securing energy supplies with obvious consequences for prices.

The challenge to meet the expected increased demand for crude oil and natural gas by new emerging economies is indeed daunting. For example, while today the United States of America consumes about 21 million barrels per day (b/d) of crude oil, China on the other hand consumes only between seven and eight million b/d. However, Chinese demand is expected to exceed 15 million b/d by 2015. This additional amount of crude oil is higher that the total current annual production of the Kingdom of Saudi Arabia or the combined annual output of the United States of America and Canada. And this is only the increased demand by China. There are many more emerging economies that will need additional supplies of hydrocarbons.

Notwithstanding this, the emergence of new promising markets is good news for hydrocarbon producers and exporters that are likely to benefit from this increased demand. Among the UNECE countries, it is the Russian Federation, the Republic of Kazakhstan, the Republic of Azerbaijan and Turkmenistan, which are adjacent or nearby to the growing markets of Asia, that are likely to be the chief direct beneficiaries of this development. However, since the market for oil is global in nature and increasingly so for natural gas, other UNECE energy exporters, such as Norway, the Kingdom of the Netherlands and Canada, are also likely to benefit from the increased demand of the emerging economies.

The importance of pipelines, ships and the liquefied natural gas (LNG) supply chain in delivering oil and natural gas to markets in an efficient and timely fashion cannot be underestimated in an ever more competitive world economy and society which demands high flexibility at reasonable cost. Moving hydrocarbons in a timely fashion and processing them to market specifications will continue to be a major challenge both for private and state-owned enterprises and for policymakers. The current technological progress and cost reductions being achieved in the LNG supply chain will increasingly contribute to the globalization of the natural gas market as well as enhancing gas deliveries to, and energy security for, both Western Europe and the United States of America.

Crude Oil Prices

Broadly speaking, crude oil prices are a function of supply-demand fundamentals, available surplus or spare oil production capacity, and geopolitical and energy security risks. The rapid growth in oil demand in recent years, particularly in Asia but also elsewhere, has meant that the growth in demand has outstripped additions to global oil production capacity. Today, demand and supply are very finely balanced. The slower expansion in production capacity in comparison to the growth in oil demand has also meant that spare production capacity has been greatly reduced. In the past, Saudi Arabia maintained significant spare oil production capacity that could be brought on stream quickly, if needed, to moderate price increases. This is not the case any more, or at least not for the time being, because Saudi Arabian spare oil production capacity is also quite constrained.

With supply and demand finely balanced and no real appreciable spare oil production capacity available, geopolitical and energy security risks have taken on added significance. Crude oil prices are constantly reacting to negative political events and energy security developments. It is estimated that 20 to 30 per cent of the price of crude oil, when prices were about US$ 70 per barrel in 2006, was attributable to geopolitical and security concerns. That translates to about US$ 15-20 per barrel. However, even if the premium due to geopolitical and security concerns was stripped out, the crude oil price would still have been above US$ 50 per barrel – this is a reflection of tight global supply and demand conditions.

It would seem that the underlying long-run energy fundamentals that prevailed in the 1970s and early 1980s have reappeared or were never really transformed. Needless to say, energy markets today are different from those that existed in the 1970s, but there are many unrelenting trends that are of concern.

Coal, Nuclear and Renewable Energy

The renewed preoccupation with energy security is refocusing the debate on the future role of coal, nuclear power and renewable energy in meeting the energy needs of UNECE countries. These energy sources are perceived to be more secure than for oil and natural gas.

Indeed, these are very interesting times for coal. Not long ago coal was viewed as having little or no future. The situation has changed dramatically in just a few years. Energy demand is increasing at a rapid rate, especially in developing countries. With high natural gas prices and supplies of coal plentiful in many countries, coal is re-emerging as a reliable and cost-effective option. Coal has the advantage that world coal reserves are large; sources of supplies are diversified; ample supplies are available from politically stable regions; world infrastructure is well developed; new supplies can be easily brought on stream; and coal can be stored.

However, coal faces many challenges, not the least of which is its environmental footprint throughout the supply chain. The greening of the coal-energy chain is vital. Existing, commercially viable clean coal technologies offer opportunities to mitigate the environmental impact of coal use at all stages of the coal cycle. Moreover, emerging new technologies (carbon capture and storage, gasification and liquefaction) could offer the potential of using coal for power generation with low or no emissions in the future and, in the longer-term, ultimately for transport. But, while the expected progress in clean coal technologies will certainly increase coal's environmental appeal, it will add both to capital and operating costs.

Since 1973, nuclear power has significantly contributed to meeting rising electricity demand in the UNECE region. However, since the early 1980s, far fewer orders for nuclear power plants have been placed, stemming in part from public concern and political debate on the possibility and consequences of accidents, on the lack of adequate methods for disposal of nuclear wastes, and over the costs of nuclear power plants themselves, including their decommissioning costs.

There are signs of a revival of interest in nuclear power, as evidenced by the decision of the Republic of Finland to move forward with the construction of a new nuclear power reactor, the decision of the United Kingdom of Great Britain and Northern Ireland to potentially resume the construction of new nuclear power plants, the continuing work on the completion of nuclear facilities in Eastern Europe (Romania, the Russian Federation and Ukraine), the rise in the resale value of existing nuclear power plants in the United States of America and ongoing work on the construction of about 27 reactors worldwide, mainly in developing countries but also in Japan. On the other hand, it should be noted that some UNECE countries, such as the Kingdom of Sweden and the Federal Republic of Germany, continue to opt against the construction of new nuclear power plants and for the phase out of current plants.

While the revival signs are there, the future prospects for nuclear power are still uncertain. Concerns about nuclear safety and the disposal of nuclear waste continue to plague the industry. Though perhaps as important are financial and economic considerations. The high upfront capital costs required and the uncertainties about the potential future liabilities associated with the nuclear fuel cycle continue to act as a major hindrance to nuclear energy investments.

Renewable sources of energy are perceived to be the most environmentally benign sources of energy and are seen as the way forward for solving many energy-related health and environmental problems. Indeed, government programmes and targets for renewables continue to be very ambitious; new initiatives, both at the regional and national levels, are being launched; direct and indirect support is being provided; and the means for financing projects are multiplying. In particular, wind and solar technologies are being rapidly developed and the installed capacity is expanding quickly.

There is no doubt that renewables will increase their market share of total energy consumption over the coming years, but they are not likely to displace, in a significant way, the use of fossil fuels over the foreseeable future. This is because of their much higher supply costs and their requirements for vast tracts of land and water surfaces. For example, between 1990 and 2004, the contribution of renewables in meeting the total primary energy requirements of EU countries rose from 4.5 per cent to 6.5 per cent, from 12.0 per cent to 14.5 per cent for electricity generation, including hydro, and from 0.8 per cent to 5.0 per cent for electricity generation, excluding hydro. The corresponding numbers for North America are from 6.5 per cent to 5.9 per cent for total primary energy, 18.6 per cent to 15.3 per cent for electricity generation, including hydro, and from 3.0 per cent to 2.4 per cent for electricity generation, excluding hydro.[1]

Thus, despite their rapid development and commercialization, the contribution of renewables in meeting the growing energy demand of the UNECE region has not appreciably increased over time and is unlikely to do so for the foreseeable future. Even the potential of hydroelectric power to contribute to increasing electricity demand is limited. The region as a whole is characterized by a state of maturing (or limits) when it comes to the development of hydroelectric power. Suitable sites are increasingly difficult to locate for hydrological reasons, competition with alternative land and water uses, and public resistance to the impact of hydro schemes on the natural environment. The Russian Federation still possesses substantial untapped resources, but these are in Eastern Siberia and are unlikely to be developed very quickly because of their remoteness and low population density. Likewise, there is still considerable potential in a number of countries in Central Asia, but their development is hampered by the same constraints as those that apply to the development of oil and gas projects.

Currently, natural gas is the fuel of choice for power generation for cost and environmental reasons. But large-scale combined lignite mining and power generating facilities remain cost competitive. The same is true for conventional coal-fired power plants where low-priced coal is readily available. However, these facilities contribute to much higher levels of environmental pollution. On the other hand, nuclear and renewables, except in special circumstances, tend to be higher cost options for power generation.

The wider the variety and types of energy sources used to generate electricity, the greater the security of electricity supplies. Over-reliance on one type or form of energy, particularly imported energy, can increase a country's vulnerability to unforeseen mishaps. A well-balanced fuel mix for generating power is the safest way for countries to ensure energy peace of mind. The choice of fuel mix for future power plant capacity can have a long-lasting and profound impact on energy import dependency, and thus on energy security considerations. Over the longer-term, nuclear power and renewable energy remain potential alternatives for electricity generation. While nuclear power may not necessarily be a desirable option for each and every country, removal of that option for all countries, as a group, would remove an important element of flexibility and diversity in energy supply and, thereby, undermine energy security for all countries.

Technology and Investment

It is difficult to predict whether there will be a significant technological breakthrough related to traditional and renewable energies in the near term. There are many barriers to energy innovation at each stage from the laboratory, through demonstration and early deployment, to widespread dissemination. However, considerable efforts are currently being expended and funds deployed by governments and the private sector to promote the development and commercialization of more advanced coal combustion and nuclear technologies, renewable energy technologies, transportation bio fuels, hybrid systems, hydrogen-based processes and carbon capture and storage technologies, that are more environmentally and publicly acceptable than many of the technologies and processes currently in use. The more new technologies are developed and commercialized, the greater will be the range of energy options available to individual consumers and countries, and the healthier will be the situation with respect to energy security.

Very large energy investments will be needed all along the energy supply chain if the expected energy demand in the UNECE and other regions of the world is to be met. Given the long lead times, the long-term nature and international character of the world energy sector particularly for hydrocarbons, as well as the relatively unstable political and economic situation in a number of hydrocarbon producing and transit countries, it is important that this investment challenge be addressed sooner rather than later.

Energy production, transport and distribution infrastructure, including pipelines, electricity grid systems, LNG terminals and ships as well as refineries, are very costly, with long payback periods, requiring huge investments. The total capital costs of Europe's first export facility for the liquefaction and shipment of natural gas, currently under construction by Statoil at Hammerfest in the Norwegian Arctic Region, is estimated at about US$ 8 billion, including the costs of the offshore natural gas production facilities. The world's first large-scale coal-fired power plant (450 megawatts (MW)) with integrated coal gasification and carbon dioxide (CO_2) capture and storage, currently under consideration by RWE Group in Germany, is expected to cost in the order of one billion Euros. To be profitable, such investments with high upfront capital costs will require relatively robust international energy prices in the future.

1.4 ENERGY SECURITY AND THE EUROPEAN UNION

Executive Summary

- The preservation of the European Union's political and economic status in a changing world energy market where bargaining power is shifting, particularly in the area of hydrocarbons, is one of most difficult strategic challenge the EU faces when it comes to energy security.

- While diversification of fuel sources, producers and transit routes is seen as the primary way of enhancing security of supply, the European Commission is also giving high priority to the completion of the internal energy market and the development of a coherent strategy for engaging with producing and transit countries outside the European Union.

- Although energy is viewed as an important driver of European Union integration, EU countries are having difficulties to forge a common energy policy and to act together over concerns about the international energy market because of a host of reasons, including differences in the mix of energy consumed domestically, different degree of access to energy resources and asymmetric interests. Furthermore, the current structure of the EU energy market does not provide sufficient incentives for private sector investment favouring a strategically diverse energy portfolio.

- On a macro-scale, the major risks for concern to EU foreign and security policymakers include the long-term potential for low-intensity conflict, war and social and political unrest in producer and transit regions, as well as terrorism and cyber attacks on critical energy infrastructure.

- The European Union therefore faces a period of debate concerning her role in the protection and preservation of energy security beyond her borders, with foreign relations and security policy increasingly overlapping with trade affairs and industrial policy.

Key Points of Energy Security for the European Union

- European Union Governments are having difficulty in determining what constitutes an appropriate appetite for energy supply risk in a world where the control of energy reserves is shifting further away from the European Union space.

- The European Union faces an underlying growth in its dependence on third countries for its energy needs. The unequal distribution of energy resources, proximities to external supplies, and differing levels of economic strength within the European Union makes the challenge of establishing a common European Union policy in this area difficult.

- Over the past twenty years, integration between energy systems has reduced inefficiencies but the liberalization of markets is creating a system with little spare capacity and margin for error. Internally, the European Union market has been focused on the optimisation of energy networks and less focused on ensuring the long-term investment required for a diversity of major projects.

- Importing energy to the European Union need not necessarily be a problem as long as the diversity of fuel types and source regions, as well as the relative vulnerability of the infrastructure used to transport and store the fuel, are in balance with the level of energy supply security that the European Union region deems to be strategically desirable.

- Higher prices in particular are changing the context of energy for the European Union in many ways, and for the countries at both ends of the oil value chain. Price volatility has become an additional issue.

- With access to energy a critical component of both economic growth and political stability, Governments everywhere (including within the European Union) are therefore increasingly examining the pros and cons of energy market intervention.

- If security of supply is or becomes uncertain (for some or all member states), or the level of security is asymmetric among the member states, the urge to implement national energy security policy by some member states, to guarantee these supplies, could well become stronger.

- Governments of major energy producing and exporting countries are establishing new bilateral relationships with major strategic energy users, in some cases reorienting geopolitical relationships around oil and gas resources.

- As the European Union's strategic energy dependency increases, so will the attention of EU nations to questions concerning the long-term stability of producer countries. The internal distribution of oil revenues has created political disputes and social unrest in various producing countries, where governance systems are weak. If the European Union fails to act in a coordinated manner on the world stage over energy issues, its influence over developments affecting the international market could erode over time and might in the end affect the European Union's energy choices in terms of markets and security.

- The tight supply/demand balance in the oil market is aggravating supply risks, and the oil industry is thus becoming more vulnerable. The shift in production away from Western Europe means there is a greater physical opportunity for attacks on infrastructure and personnel, and as the value of oil revenues rises, greater damage can be inflicted by such attacks. Furthermore, it cannot be discounted that over time the European Union's own energy systems could be subjected to attack.

- The European Commission's Green Paper "A European Strategy for Sustainable, Competitive and Secure Energy" recognises that the margins for manoeuvre are smaller in terms of energy supply than in terms of energy demand. Policies for controlling demand are, therefore, a priority.

- Financial strength, market power and capital market access are crucial elements in helping to mitigate the European Union's security of supply risks. Utilities in the European Union with these features tend to be the biggest and most diversified companies. Additional supply concerns over time could drive further consolidation in the European energy market.

- Although the more developed European Union countries are generally successful in attracting private sources of finance, the situation in the less developed countries is challenging and requires improvement through the establishment of a more predictable legal, policy and regulatory context and a more favourable investment climate.

- If European Union Governments wish to encourage longer-term investments in the interests of ensuring incremental, interconnected or replacement infrastructure, and to ensure a more diversified energy portfolio, policymakers must consider how to effectively stimulate private sector investment, both to develop new technologies and the new capacities required to meet growing demand.

- Industry and government need to liaise and cooperate on a continuing basis, and allocate responsibilities between themselves within strategic, integrated risk management frameworks. Integrated risk management frameworks should include mechanisms for risk governance, risk assessment, risk quantification, monitoring and reporting, and risk control optimization, i.e. a system that identifies and controls short-term and long-term risks from all sources.

Europe's New Era of Energy Insecurity

As European Union Governments have come to view energy security as a political priority so too have they recognised that absolute energy supply security is not attainable, a condition that is unacceptably expensive to guarantee. Energy security is a fundamental economic trade-off that governments and societies must make, between the cost of delivering different levels of energy security with their ability and willingness to pay for it.

The strategic challenges facing European Union policymakers are complicated by the fact that the risks impacting global energy systems are occurring with greater frequency and volatility, within an energy market that is increasingly integrated, capacity-constrained and competitive.

Governments of the European Union are therefore now considering what constitutes an appropriate appetite for energy supply risk in a world where the control of energy reserves is shifting further and further away from the European region.

The difficulty for the European Union is essentially how to preserve its political and economic status in a changing energy world with the bargaining power shifting to energy producers and exporters. Like the United States of America, the Republic of India or the People's Republic of China, the European Union faces an underlying growth in its dependence on third countries for its energy needs. The unequal distribution of energy resources within the European Union countries, the different proximity to external supplies, and differing levels of economic strength within the European Union makes the challenge of establishing a common European Union policy in this area difficult.

Furthermore, over the past twenty years, integration between energy systems has reduced inefficiencies but the liberalization of markets has reduced spare capacity with little margin for error. Although initially bringing price benefits to consumers, this trend has increasingly exposed the European Union supply chains to impacts that can quickly travel throughout the system.

Internally, the European Union market has delivered in terms of the optimisation of energy systems and networks, but not on the long-term infrastructure investment required to develop a diversified portfolio that adds incremental capacity in a timely manner, and which mitigates against strategic risks.

For example, as long as spare capacity is thin along, or at any point in, the Europe Union's energy supply chain, disturbances will tend to generate price spikes. And even when adequate spare capacity is restored, most future supply will come on stream in significant lumps, with the potential to cause downward price spirals if not adequately addressed.

The European carbon market has been a great success for traders, but less so from an environmental perspective: high oil prices has encouraged utilities to burn more coal, in turn driving carbon prices up. Moreover, the carbon market as currently configured offers investors no confidence over the sort of timescales necessary to drive infrastructure investment.

Private energy utilities within Europe seek profit and, given strong market volatility, they will seek a price on delivering new infrastructure in a volatile market. On the other hand, most European Union utilities now operate with a cross-border perspective and, taking the European Union as a whole, can serve to stabilise physical cross-border supply deficiencies increasingly well.

Timely investment for European Union projects is essential, but as long as there is uncertainty as to what Governments might do, investment can be stalled, and the longer this occurs, the more acute the problem.

While European Union concerns are no longer simply characterized by the risk of oil supply disruption from OPEC producers, this remains a significant concern in a crude market with tight capacity and in which the non-OPEC share is declining steadily. In itself, however, importing energy to the European Union need not necessarily be a problem as long as the diversity of fuel types and their source regions, and the relative vulnerability of the infrastructure used to transport and store the fuel, are in balance with the level of energy supply security that the European Union region deems to be strategically desirable.

Implications of the Oil Market

Geopolitical competition for oil and gas has the potential to become a destabilising force in many producer regions with undiversified economies. Increasingly, bilateral energy relations are becoming a top priority for many consumer countries concerned about rising oil prices and their desire to limit the proportion of their import needs that are exposed to the spot market.

Higher prices in particular are changing the context of energy for Europe in many ways, and for the countries on both ends of the oil value chain, price volatility has become an additional issue. Asymmetric interests in production levels and prices among producing and consuming countries is contributing to oil market instability. Looking further ahead, domestic economic pressures in the producer countries may become a more and more forceful driver of oil policies that could potentially undermine a cooperative oil policy stance among the main producer countries and cause additional instability, affecting the Europe Union oil market.

With access to energy being a critical component of both economic growth and social welfare, governments, including many within the European Union, are taking a fresh look at energy market intervention, with the pendulum swinging back towards statism in recent years.

Any statist trend clearly has major implications for a unified, liberalized European Union approach to energy, but also for how the European Union approaches its relationship with the rest of the world as the world moves into an era where global energy systems are increasingly not just a function of trade relations, but of foreign relations too.

A further issue that complicates the European Union's strategic thinking (as well as the investment decisions of the oil industry itself) is the fact that oil prices have become increasingly divergent from market fundamentals. Stock movements have traditionally charted oil prices, but since the Iraq War there has been significant divergence, which suggests the oil price is being affected by factors other than just fundamental supply/demand economics. Some of the divergence is due to the increasing security risks and lower spare capacity. Pipeline attacks in the Republic of Iraq have produced significant mini-spikes in the oil price, but they have also contributed to a general surge upwards.

Fortunately for the European Union, its economies are less dependent on oil than they have been historically and government stocks can act as a buffer to any shocks if they occur. However, the European Union is becoming more heavily dependent on gas imports, and the price of gas remains coupled with that of oil – price rises in gas are additionally reflected in power price rises for increasing numbers of consumers, in turn affecting European Union's competitiveness.

Box 1.4.1 Key Global Energy Statistics

KEY GLOBAL ENERGY STATISTICS

According to the International Energy Agency, global demand for oil will rise by 40 per cent by 2030. Mature economies are becoming increasingly dependent on oil imports as production from most non-OPEC countries is declining rapidly. By 2030, the main producer countries in the Middle East and North Africa will contribute 44 per cent of the world supplies compared with about 35 per cent today. By 2030, the Russian Federation and the Central Asian states will contribute a further 15 per cent of world supply, on a par with Saudi Arabia, and exceeding North American output. Significantly, they will supply 35 per cent of the *new* demand that the world will require.

In 2006, total consumption of primary energy in the world reached almost 11 billion tonnes of oil equivalent (toe), in comparison with only 8 billion in 1990. Global energy consumption is expected to increase by almost 52 per cent of its 2003 level by 2030. Between 2000 and 2020, global demand for energy is expected to increase by approximately 2.2 per cent per year. China's energy demand is forecast to experience an average annual growth of 4.7 per cent over this period.

In terms of market shares, oil is the most widely used energy source, amounting to 35 per cent, due to the importance of the transport sector where it still has no serious competitors. It is followed by coal and natural gas, with almost equal shares (23 per cent and 24 per cent respectively). Nuclear energy represents just under 7 per cent of global consumption. Hydroelectricity and other renewable energy sources represent around 11 per cent. Fossil fuels comprise 82 per cent of global energy demand.

In geographical terms, the distribution of energy consumption shows that North America is the largest energy consumer, accounting for 29 per cent. The European Union consumes 17 per cent. Developing countries in Asia represent over 20 per cent, 11 per cent of which is accounted for by China and 4 per cent by India. Latin America consumes only a little more than 6 per cent of world energy and Africa uses less than 3 per cent.[2]

Market stability is clearly an important goal for the European Union which, as a major consuming bloc, has a critical role to play in addressing immediate global needs within the energy market, including:

- The need for more investment, especially in refining;
- Improving the quality of oil market information and the transparency of data;
- Cooperation and dialogue between producers and consumers; and
- The increased burden of higher energy prices on poor, oil importing countries.

Market stability, well-informed and flexible government policies, and clear signals for investment to address strategic regional energy supply risks must therefore form the bedrock of the European Union's attempts to optimize its long-term energy supply security.

On top of these political economy issues, the European Union must also consider what strategic stance it should take with regard to foreign and security relations in external producer and transit regions.

Box 1.4.2 Ten Key Principles of Energy Security

TEN KEY PRINCIPLES OF ENERGY SECURITY

- Diversification of energy supply sources is the starting point for energy security.
- There is only one oil market.
- A "security margin" consisting of spare capacity, emergency stocks and redundancy in critical infrastructure is important.
- Relying on flexible markets and avoiding the temptation to micromanage them can minimize long-term damage.
- Understand the importance of mutual interdependence among companies and governments at all levels.
- Foster relationships between suppliers and consumers in recognition of mutual interdependence.
- Create a proactive security framework that involves both producers and consumers.
- Provide good quality information to the public before, during and after a problem occurs.
- Invest regularly in technological change within the industry.
- Commit to research, development and innovation for longer-term energy transitions.[3]

The European Union's Energy Security in the Context of Foreign Relations

The European Commission made recommendations in the year 2000 to include energy issues more prominently in external trade and foreign and security policy-making, based on the fact that the dependency on imported energies will increase substantially in the coming decades (COM 769 final, 2000) and that the uninterrupted flow of energy will mainly depend on the political and economic stability of the producer regions.

Due to the growing energy import dependency of other main consumer regions, such as the United States, India, China and other Asian countries, energy relations for the European Union countries as well as the EU as a whole may become increasingly politicized. Governments of major energy-producing countries are increasingly establishing new bilateral relationships with major strategic energy users, reorienting geopolitical relationships around oil and gas resources.

As Europe's strategic dependency increases, so will the attention paid by European nations on questions concerning the long-term stability of producer countries. The internal distribution of oil revenues has created political disputes in various producing countries, where governance systems are weak.
Given the position of elites in society and the large amounts of capital involved in oil and gas production and in government revenues and expenditures, producing countries tend to be more vulnerable to corruption practices. Unstable regions can easily become further destabilised when rising oil and gas revenues, due to higher prices, are not adequately shared among all segments of society, thus contributing to accentuating social, economic and political instability. Consuming countries are therefore increasingly concerned with preventing such situations from developing.

The risk for the European Union is that when it attempts to act geopolitically on the world stage over energy issues it often does so as an uncoordinated player, unable to get consensus among its own members, and thereby having less influence over developments in the international energy market. In such a scenario, energy choices made by the United States and, increasingly, those made by the expanding economies of China and India, could end up in constraining the European Union's own energy choices in terms of markets and security.

The strategic energy risks facing the European Union region were brought home recently by the Russian-Ukraine and Russian-Belarus events in early 2006 and 2007 over gas and oil deliveries respectively. These not only had an impact on gas deliveries within Western Europe, but they also resulted in some confusion about how to interpret and respond to the events.

The effectiveness of the European Union's strategic energy policy tools thus depends not only on the EU's ability to employ domestic energy assets, on technical and operational factors, on transportation and import facilities, on the investment climate and the availability of foreign oil and gas supplies, but also on the geopolitical setting in which these policies must perform. Given the dynamic developments in international political and economic relations, a static approach to European Union energy security – whether defined at the national, regional or international level – could be detrimental to the region.

Box 1.4.3 The Trend Towards Statism

THE TREND TOWARDS STATISM

The motivation for statism and the form that it takes varies from country to country. In India, for example, the government has adopted statism as a default position for electric power, because private investors have been reluctant to invest, given the current political and economic environment. In Latin America this issue is country dependent, but the recent history of poor investment results has discouraged all but the most committed private-sector players.

Strong state-owned companies sometimes fill the void left by reluctant private-sector investors. In Russia, the shift back toward statism has favoured oil and gas state-own companies.

Control over gas pipelines and exports to the European Union is only one example of the strong role that the Russian Government is playing. In the Middle East, state control of oil and gas resources prevails; however, the region is opening up to private investors in the power sector. In Europe, strong regulatory intervention in the power sector, designed to steer energy investments towards lower-carbon fuels and more open, liberalized markets, is sending confusing signals to investors and consumers alike. Governments of oil-exporting countries also maintain their power over natural resources, not always sharing the financial rewards of extraction with those most affected by the exploration and production activities.[4]

Physical Security Issues and Associated Challenges for the European Union

This raises the fundamental question of how the European Union would, or could, act collectively to intentional restrictions by third parties to its energy supplies. The North Atlantic Treaty Organisation's (NATO) Article 4 states that, "The Parties will consult together whenever, in the opinion of any of them, the territorial integrity, political independence or security of any of the Parties is threatened." The 1999 NATO Strategic Concept, Paragraph 24, is more specific: "Alliance security must also take account of the global context. Alliance security interests can be affected by risks of a wider nature, including acts of terrorism and by the disruption of the flow of vital resources." The tight supply/demand balance in the oil market is now aggravating supply risks, and the oil industry is thus becoming more vulnerable to attack. The shift in production to non-EU countries and the lengthening of transportation routes means there is a greater physical opportunity for attacks, and as the value of oil revenues rises, greater damage can be caused by any such attacks. Insurgent/terrorist groups will recognise the vulnerability and potentially target increasingly oil installations where possible.

Ultimately, prices are always a question of supply and demand, while security and geopolitical risks are concerns about future demand which can add a premium to the price. The current geopolitical and security risk premium is exaggerated by tight capacity and it is difficult to isolate specific security threats within the crude oil price. However, whilst quantifying the risk premium is subjective, some analysts calculate that up to 20 - 30 per cent of the total price in the past few years constituted security and geopolitical risk.[5]

The geopolitical risk premium pales in comparison with the impact of a significant strike on oil infrastructure, though. Big oil price movements seemingly only occur when very significant events happen, impacting for lengthy periods on the market, and these are still rare. However, it should be remembered that explicit threats to key components of the world oil system are now regarded as being credible and made by organizations with the likely capacity to execute the threats.

The energy sector is vulnerable to attack in a constellation of ways, from initial geophysical survey teams at risk because they do not know an area, to rebels simply stealing equipment and infrastructure from entrenched facilities. Approximately 1,000 significant events occur each year and in recent years more than half of these occurred in the Republic of Iraq and the Federal Republic of Nigeria. Tackling these requires close liaison between companies and host governments across the whole value chain.

It would be tenuous to say that insurgents/terrorists acting in different countries have shared much information on how to attack energy facilities; attacks so far are primarily based on local knowledge. However, insurgents in Iraq and the Republic of Colombia clearly have had insider information with respect to facilities because they know how and where to attack for greatest effect. Therefore, it cannot be discounted that over time Western Europe's energy systems will not be subjected to some type of insider penetration attempts.

Terrorist impacts on international energy infrastructures, however, would currently only be very significant for European Union countries in the case of a prolonged disruption to energy supply routes, or indeed where major damage to key producer terminals was coincident with capacity loss arising from some other event, such as a hurricane disrupting production elsewhere in the world. In fact, the greatest impact and danger is from the cumulative impact of a number of events, whether these are due to terrorism or natural causes, occurring simultaneously or close in time. This could, if significant, trip the world financial markets for a period and, whilst European Union consumer economies could clearly suffer, the impact on developing world economies would probably be greater and more destabilizing.

Within the European Union, industrial control systems in the energy sector are becoming increasingly vulnerable to cyber attack owing to the move away from customized applications to the adoption of standard information technologies (IT). Aggressive penetration of IT systems is now capable of infecting global networks within minutes with core operating systems frequently being attacked.

The types of threat that European Union energy companies are facing include the standard ones – unauthorised control, loss of integrity and loss of confidentiality – but also, significantly, denial of service, inhibiting their ability to deliver supplies. The obvious impact is the financial consequences of disruption to production or operations, but there are other consequences, including: environmental impacts, health and safety concerns, inability to honour client commitments, loss of reputation, non-compliance with regulations, threat to operating licence, and physical damage to infrastructure.

This highlights the additional challenges facing the European Union – precisely which aspects of the energy system need protecting and from whom? Is it: exploration and drilling operations? Production, generation and refinery operations? Pipelines and transmission lines? Transportation, or downstream points of delivery? Essentially, energy infrastructure is extremely varied as are the associated protective requirements, which are frequently long-term and large.

A second related issue is which locations require protection. Are these within the developed world or in the developing world where consideration might have to be given to conflict zones, post-conflict or high-terrorism-threat locations, and/or politically-challenging and higher-criminality environments? Is the area hostile, remote or urban? There is a massive array of locations with a host of different challenges in which the European Union's energy interests operate globally.

A further issue refers to the nature of the threats from which protection is needed. Are these threats: situational events arising from conflict zones; local insurgencies; international terrorism; siphoning from pipelines; sabotage, vandalism and theft; protest; or corruption, fraud and semi-official threats? At a more macro level, the major risks going forward that the European Union's foreign and security policymakers need to be concerned with, include the long-term potential for low-intensity conflict; social and political unrest in a number of regions or countries; and terrorism.

A potential answer is to bring producer and consumer governments and companies together to examine past experiences from all over the world, and to focus on security and contingency planning in the context of an integrated approach to energy sector risk mitigation. For now, however, the European Union is just beginning the debate on how to develop a collective approach to the strategic problems that it faces.

Box 1.4.4 Trends in Global Energy Security Threats

TRENDS IN GLOBAL ENERGY SECURITY THREATS

Past, Present and Future Energy Threats

- 1960s-1980s: Energy infrastructure security threats mainly from Soviet-bloc supported Communist insurgencies and civil wars.
- 1990s: Post Soviet-era led to collapse of insurgencies, or their ability to fund operations through other mechanisms (*e.g.,* drug trade).
- Mid-1990s-Present: Growth in global Islamic militant threat to Western energy company infrastructure and personnel.
- The potential for transnational conflict, civil war and low-intensity conflict will remain a pre-eminent energy security threat, especially in the Gulf.
- The future: Banditry/crime/attacks in unstable operating environments; Islamic militant terrorism against Western energy interests.

Global Energy Security Threat Mitigation

- Tight world oil supply/demand balance means that disruptions or perceptions of threats will affect markets more than in the past.
- Expanded natural gas utilization will increase the need for security planning and transnational risk mitigation efforts.
- Incorporation of a broader range of stakeholders needed upfront in projects.
- Closer inter-governmental and government-company cooperation needed to ensure energy security and unrestricted energy flows within world markets.
- Learning from past security experiences and transferring this knowledge to operations in other parts of the world is critical to employing the best risk mitigation practices possible.[6]

Towards a Common European Union Energy Policy

According to the forecasts of the International Energy Agency (IEA), Europe will have to import almost 70 per cent of its energy needs in 2030, compared to 50 per cent at the moment. Thus, the European Union will be 90 per cent dependent on imported oil and 70 per cent on imported gas.[7] However, the Europe Union's dependence on imports is not a strategic weakness in itself. Energy security is not aimed at maximising energy self-sufficiency or minimising dependency but rather reducing the risks related to dependency on imported energy.

The European Commission has also viewed the European Union's energy market with a degree of concern. Two reports presented by the European Commission in 2005 concluded that:

- gas and electricity markets are still too concentrated;
- national markets remain too segmented;
- historical operators continue by and large to influence the market and price levels; and
- there is a severe lack of transparency within the markets and in the setting of prices.

Judicial activity by the European Commission's Competition Directorate General in support of the energy liberalization agenda indicates a will to tackle these barriers to fully open markets, but while competition is seen as vital many EU member states have yet to put liberalization fully into practice.

While diversification of fuel sources, producers and transit routes is seen as the primary way of enhancing security of supply, the European Commission's Energy Green Paper (SEC (2006) 317/2) prioritizes the completion of the internal energy market, highlighting the need for market information and transparency and calling for effective unbundling and consistency of regulatory frameworks, as well as for a coherent strategy to engage with producing and transit countries outside the EU.
The European Commission believes that the EU can achieve collective influence in foreign policy that individual states cannot. According to the Commission, adopting a strategy on the security of supplies at the Community level is essential in order to limit the risks incurred, in as much as the national energy policies of EU member states impact on each other. However, concerns exist about the possibility of statism derailing attempts to further the European Union's liberalization agenda. If security of supply is or becomes uncertain (for some or all member states), or the level of security is asymmetric among the member states, the urge to implement national energy security policy by certain member states, to guarantee these supplies, might well become stronger.

The European Union countries also have different concerns when it comes to energy dependence. Perceptions are influenced by a country's geographical position (the Baltic States, the Republic of Poland and Finland's proximity to the Russian Federation), a flourishing economy (Ireland, the Republic of Austria), the state's power to make energy choices (France), the importance of the oil sector (the Republic of Italy) or the gas sector (the Kingdom of Spain) in overall energy consumption, as well as progress in measures to liberalize the energy market (the United Kingdom, Germany or Benelux). To the extent that member states find it necessary to forge national security of supply policies at the level of national foreign policy-making, unilateral statist strategies to deal with supply concerns will not only interfere with EU energy policies but could have negative effects on the development of EU foreign and security policies. Considering the external energy dependency of the EU and given the internal market, it may be that the EU has no other alternative but to develop a coherent energy security policy that also addresses the fundamental asymmetry in exposure among the member states.

Energy is thus viewed by the Commission as an important driver of further European integration. The Energy Green Paper cites a range of reasons for a common EU approach to energy: to equip the EU to play a full role in global markets; to improve sustainability in the EU and globally; to improve the functioning of the internal market; to improve stability in the EU and neighbouring markets; and to reflect the strategic role of energy in achieving other political objectives.[8]

Graphs sourced from the European Commission's Green Paper clearly highlight why the Commission is now exhorting policymakers to address the European Union's strategic energy security agenda with considerable urgency (Figures 1.4.1, 1.4.2 and 1.4.3).

Figure 1.4.1 Looking Ahead – European Union 25 Energy Import Dependency (per cent)

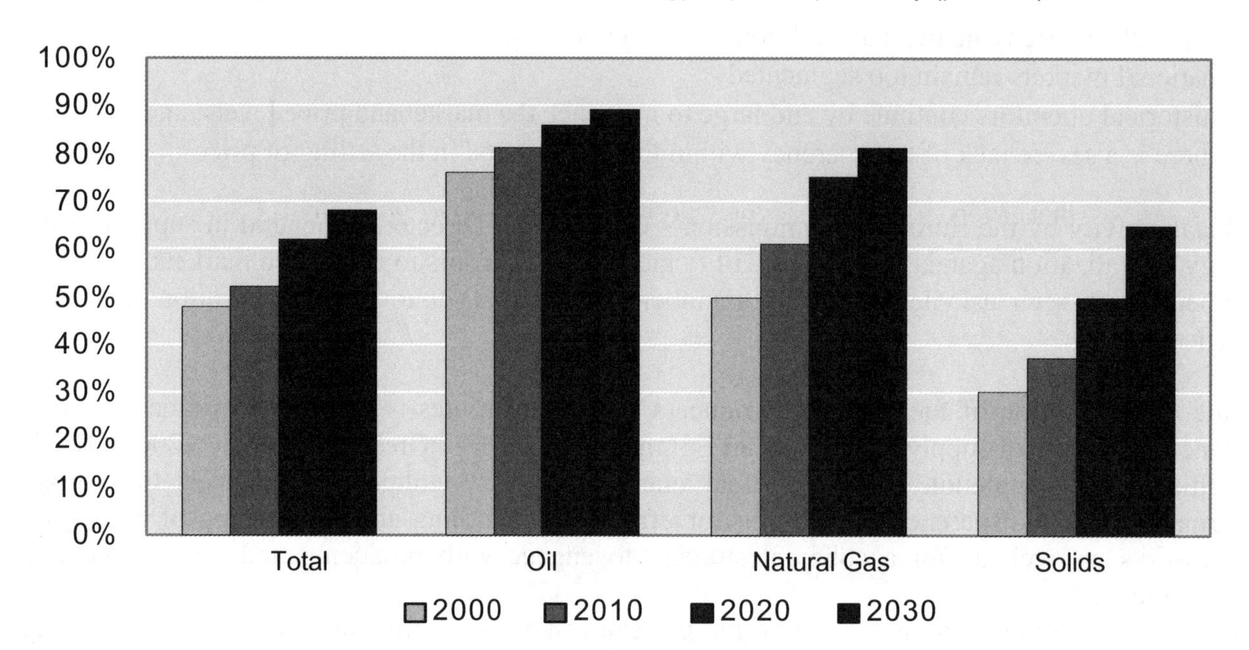

Source: European Commission Green Paper "A European Strategy for Sustainable, Competitive and Secure Energy"

Figure 1.4.2 EU Final Energy Demand (mtoe)

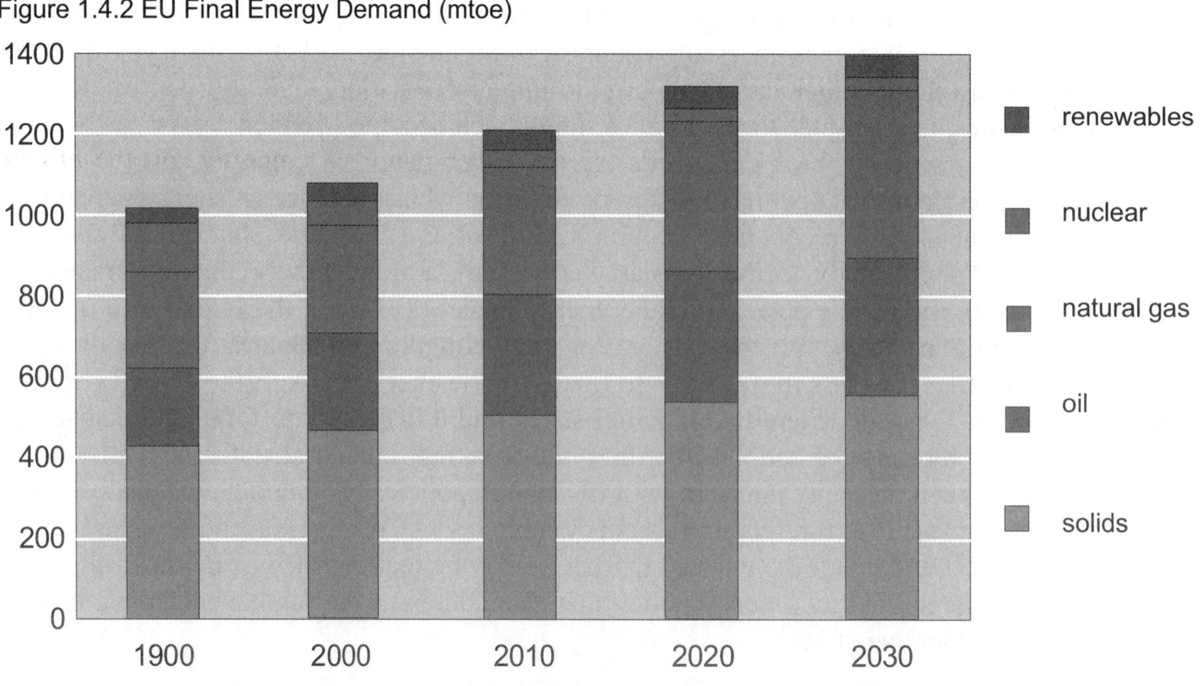

Source: European Commission Green Paper "A European Strategy for Sustainable, Competitive and Secure Energy"

Figure 1.4.3 The Competitiveness Challenge

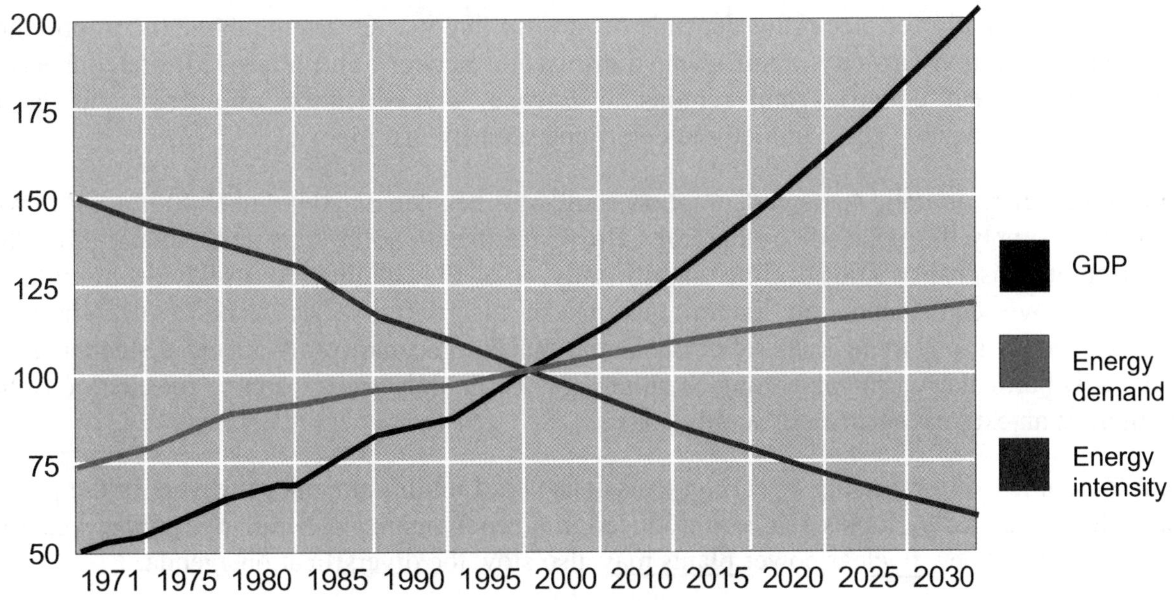

Source: European Commission Green Paper “A European Strategy for Sustainable, Competitive and Secure Energy”

The European Commission’s Energy Green Paper

The European Commission’s Green Paper “A European Strategy for Sustainable, Competitive and Secure Energy” recognises that the margins for manoeuvre are smaller in terms of energy supply than in terms of energy demand. Policies for controlling demand are, therefore, a priority and must be based on the principles of energy efficiency and effectiveness as well as the use of taxation as an instrument for steering demand towards more controlled consumption, which is more respectful of the environment. Supply-related measures must aim to develop national energy supplies which double the proportion of renewable energy contributing to total energy use by 2010, provide a strengthened system for strategic stocks and the construction of new infrastructure for transporting energy.

Key points from the Green Paper include:

“As far as oil is concerned, the EU depends, as any other major oil importer, on the effective functioning of a global oil market. EU security of oil supply, whether seen as protection against disruption of supplies or against excessive prices, thus has to be measured against the global market.”

The Commission is confident with regard to regional gas supplies for the coming 20 to 25 years:

“Over and above the huge Russian reserves, Norway, North Africa, Nigeria, Middle East and the Caspian Basin all hold large gas reserves in development or waiting to be commercialised. The proximity of the established EU market makes the EU a very attractive customer for these countries/regions.”

The Green Paper further states the following with regard to electricity:

"The security of supply for electricity depends on several factors. Over and above the importance of sufficient generation capacity, a sufficient transmission network and improved interconnection between the different countries and regions in Europe, diversification of energy sources in electricity generation is the key parameter to determine security of supply.

"A drastic reduction in solid fuels consumption could still be seen as compatible with a high level of security of supply if "replaced" by nuclear. However, if both solid fuels and nuclear, albeit for very different reasons, would decline significantly, the EU might see itself faced with an unacceptable over-dependence on natural gas in the electricity sector, even if the share of renewables would grow more than reflected in the baseline scenario. A strong dependence on natural gas is particularly critical as long as natural gas is the only energy source for electricity for which there is an external security of supply concern."

A point to note is that the pricing of carbon emissions could inhibit attempts to diversify the power sector, unless clean technologies can be introduced in a timely manner. Similarly, public concerns over the building of new nuclear power plants may also slow the diversification agenda.

"Nuclear power may satisfy the need to diversify and at the same time reduce the import dependency and limit the emission of greenhouse gases. The current plans to dismantle nuclear capacity will begin to have a serious impact on the EU generating capacity from 2015 onwards. At the moment, nuclear power is not generally accepted by large parts of the European public due to the problem of nuclear waste disposal. However, alternatives for nuclear power might be very hard to find in the short term without further increasing the import dependency, because the contribution of more sustainable energy will only slowly increase."

Regulatory and market reform, including price flexibility and the removal of existing barriers to trade are clearly essential for the operation of efficient energy systems and best suited to attracting the much needed capital investment. Although the more developed countries of the European Union are generally successful in attracting private sources of finance, the situation in some countries of Central and Southeast Europe is challenging and requires improvement through the establishment of a predictable context and a favourable investment climate.

In the past, access to energy was not an issue for the new members of the European Union located in Central and Southeast Europe. Heat and power were considered free goods. There was no metering, no payment and no disconnections. Energy was delivered below cost and environmental impacts were not seriously considered. However, in the post- transition period, and more recently as members of the European Union, there has been an increasing awareness that energy efficiency is their countries' most abundant natural resource.

The situation in those countries is also changing for a variety of reasons, including: rising fossil fuel prices, the need for a competitive domestic industry, and import substitution.

The Importance of Capital Markets in Support of European Energy Security

Within the Europe Union, and indeed globally, there is clearly a pressing need for infrastructure investment to assure security of supply. Importantly, however, there is no appreciable constraint on the availability of financing to meet these needs. The issue is rather the nature of frameworks that need to be put in place to create the right conditions to meet investment objectives. If left entirely to the market, and/or where future governmental frameworks are uncertain, investments tend to be short term. Markets predictably gravitate towards the simplest investment focussed on the short term, and within Europe at the moment this focus is on gas.

Box 1.4.5 The European Commission's Green Paper Proposals

THE EUROPEAN COMMISSION'S GREEN PAPER PROPOSALS

Priority areas for possible common action:

Moving towards a fully competitive internal market; solidarity amongst members in terms of internal energy supply policy; a diverse, efficient and sustainable energy mix; an integrated approach to tackling climate change, including energy efficiency measures and renewable and low-carbon energy production; a strategic approach to technology and innovation; and a coherent external energy policy.

Proposals for discussion with regard to internal energy supply policy:

Full development of internal gas and electricity markets; a strategic review of the European Union energy mix; a European energy supply observatory; an action plan on energy efficiency; a road map for renewable energy; a strategic energy technology plan; and a common regulatory framework for the EU's internal gas and electricity markets.

Proposals for discussion with regard to solidarity in an internal market:

Improved network security; a new mechanism to ensure solidarity and assistance in the event of damage to infrastructure; a priority list for new infrastructure; a rapid and coordinated reaction to emergency external energy supply events; a review of the existing Community approach to oil and gas stocks; and common security standards to protect essential energy infrastructure.

Proposals for discussion with regard to an external energy supply policy:

A comprehensive and strategic EU energy review; an international agreement on energy efficiency; a pan-European 'Energy Community Treaty'; rapid ratification of the EU Energy Charter Treaty and conclusion of the Transit Protocol; limiting the EU's growing dependence on imported fuels; and leadership on climate change and the search for energy solutions.[8]

If the Governments of the European Union therefore wish to encourage longer-term investments in the interests of ensuring incremental, interconnected or replacement infrastructure, and to ensure a more diversified fuel portfolio, policymakers must consider how to effectively stimulate private-sector investment, both to develop new sources of technology and the new capacities required to meet growing demand.

There are three constraints on investment in new technologies in the European Union at the moment. Firstly, there is an unclear planning and approval process for most new technologies – the length of time, outcome and regulatory process are uncertain, and projects are liable to be stalled.

Secondly, there is a lack of a clear policy framework with respect to incentives, carbon allocation etc, with even less clarity with respect to governmental willingness to backstop certain risks in liberalized markets. The current absence of an EU-wide policy framework greatly inhibits progress in this area.

Thirdly, there is the Europe Union's market structure: government control over some energy markets adds political risk; the lack of interconnection leads to price volatility in 'island' markets; and emerging national champions create barriers to entry.

Resolution of these policy issues could conceivably emerge as a result of the European Commission's Green Paper consultation, but the core requirements needed to stimulate private investment will, however, need to incorporate clear EU-wide joint initiatives to promote new technologies, a clearly-defined carbon-permitting framework for Phase II and III, carbon used to create an incentive revenue stream for new/desired technologies, and government willingness to backstop some risks (especially with respect to any new future nuclear generation).

With appropriate European Union policy and market frameworks in place, financing will readily become available. However, in the context of this new energy era financing needs to be regarded as an essential output of policy and market frameworks. Policymakers cannot therefore assume that investment will automatically be forthcoming – they must fully understand the political and market criteria that will make long-term projects commercially and practicably deliverable.

Fundamentally, policymakers need to ask what objectives they have for the investments they would like to see. They do not need to invest themselves, but they do need to create the necessary investment conditions. This will be a key test of how consensual any future European Union-wide energy policy can be.

The necessary flows of investment into future European Union energy developments also require an examination of how this will affect the main corporate bodies within Europe. As noted previously, the market is cautious about infrastructure investment because of the potential for political/regulatory change and, since politicians' timeframes tend to be shorter than the required investment horizons, ways and means will have to be found to bring greater certainty to policy and legislative measures. However, companies will never be able to avoid all political risk and those best suited to absorbing it in the Europe Union are clearly those with the largest scale and diversity, the highly-rated companies with greatest access to capital.

In terms of the ability to raise incremental capital for energy-related projects, supply risk will not be a significant direct mid-term threat to European Union utility credit ratings, but over time security of supply concerns will drive merger and acquisition activity (M&A) and investment in a more diversified corporate portfolio in order to mitigate risk exposure, which in turn can impact on ratings and the ability of a utility to leverage subsequent finance. The ratings of utilities based in the EU-15 show a downward trend owing to liberalization and subsequent M&A activity. In contrast, the ratings of Eastern European utilities are generally on an improving trajectory, albeit from a low base.[9]

Financial strength, market power and capital market access are crucial elements, then, in helping to mitigate the European Union's security of supply risks. European Union utilities with these features are the biggest and most diversified companies. Additional supply concerns over time is likely to drive further consolidation in the European Union.

In summary, there are broadly three key challenges for European Union policymakers. The first is security of supply – generation and sourcing diversity are vital in countering the risks posed by increasing import dependency. The second is liberalization, which boosts efficiency, but creates market risk and, therefore, regulatory certainty and a clear competition framework are vital to encourage investment. The final challenge is climate change policy, which favours carbon-free generation – this currently has policy uncertainties associated with it that are delaying investment, and additionally the pricing of carbon could run counter to diversification.

Policy dexterity will thus be key to avoiding conflicting goals and uncertainties within the Europe Union, and this requires an appropriate model for strategic risk assessment, not least to determine how the best mix of public-private sector risk mitigation techniques can be employed in particular circumstances for the optimum outcome.

Optimising Outcomes: Toward an Integrated Risk Assessment and Mitigation Framework

The risk assessment and mitigation process developed and refined over many years by both public and private bodies is highly applicable to managing strategic issues of energy security within the

European Union. As discussed earlier, energy security risk mitigation for the European Union essentially centres on diversifying supply sources, energy sources/utilisation, and supply routes/methods. It also means multilateral engagement, energy efficiency and reduced consumption, investment in technological development, improved security at home and overseas, and creating strategic reserves.

Private sector energy companies have risk profiles of incredible breadth and depth, with additional analysis provided by the financial institutions that support them via the flow of strategic investment capital. Clearly it is crucial for energy companies to identify and examine all risks in order to ensure they can secure their own energy supplies for sale, but also it is necessary in order to understand how the risks interrelate with one another and to overcome the 'silo' approach that historically has plagued both the public and private sector.

Effective coordinated risk management therefore means generating 'content' and 'process'. Content means redesigning risk assessments, integrating information from all risk disciplines, and ensuring that assimilated risk information is always 'live' and contemporary.

Evaluation of developing risk content means creating risk maps, which identify the threat-level for every risk, as well as how likely it is to occur. Although it is important to quantify all risks where possible, this risk-mapping approach also requires private sector organizations and governments to incorporate unquantifiable risks, too. Once organizations have this information, they think about how to incorporate it into their day-to-day risk management and practice.

The global energy risk environment is highly dynamic, and increasingly so, such that specific political, regulatory or corporate measures need to be able to change and respond within the context of a flexible integrated risk management framework. Such frameworks must include mechanisms for risk governance; risk assessment; risk quantification, monitoring and reporting; and risk control optimization, i.e. a system that identifies and controls short-term, and long-term risks from all sources.
The steps required to optimally limit the frequency and magnitude of energy security risks may be regarded as a five stage process, but one which is carried out on a continual rather than occasional basis, and one which examines the full range of risk scenarios from the macro down to the micro scale:[10]

1. Governments must determine the public's tolerance to risk, and also understand the critical aspects of the energy supply chain.

2. Identify threats, which involves a comprehensive and proactive appraisal rather than simply being reactive.

3. Evaluate the likelihood of particular events and their economic costs. Although it is very difficult to quantify impact or probability, an imperfect calculation is better than none because some cost-benefit analysis is necessary.

4. Apply the criteria for risk tolerance established when assessing society's appetite for risk at the beginning of the process, and then identify which risks are acceptable, which must be reduced and which must be avoided altogether.

5. Mitigate those risks, and undertake crisis and continuity planning. Crisis management and continuity planning means scenario planning, identifying early warning signals, creating monitoring, reporting and escalating mechanisms, and designating responsibilities and accountability. Moreover, continuity plans need to be flexible and dynamic to accommodate change in the geopolitical climate as well in key public and private sector personnel, whether these be analysts or decision makers.

This last stage is really the crucial test for decision makers, since without effective execution of risk mitigation measures the previous steps in the process become largely pointless.

Implementation means having strong commitment and leadership from government and industry. It means establishing processes for implementation, maintenance and review. Industry and government need to liaise and cooperate on a continuing basis, and allocate responsibilities between themselves. There are often strong lines of communication between government and industry during intermittent policy review periods, but these are frequently framed around particular policy questions rather than contributing to an overarching, 'rolling' risk model.

There is also a widely held perception that (in liberalized markets, at least) the private sector only approaches governments when it wants something, and governments are thus not particularly responsive to it. To engender the necessary integrated approach to risk assessment and mitigation the market must demonstrate to government that it has mutual concerns, that it is willing to contribute to solving the problem, and that it simply wishes to discuss the kind of mechanisms and allocations of responsibilities necessary to do this for mutual long-term benefit.

What is now required is the creation of an effective risk management process for optimising the capacities of the public sector, the private sector, and NGOs to appropriately assess strategic energy security risks. Such a framework would support the development of straightforward mechanisms that allow the deployment of risk mitigation solutions where responsibility may be clearly defined and shared by the public and private sectors.

1.5 AN EASTERN PERSPECTIVE OF ENERGY SECURITY

Executive Summary

At present, renewable and new alternative energy sources deliver about 13 per cent of the world's energy needs - 2.2 per cent accounted for by hydro, 11 per cent by biomass and waste, and 0.5 per cent by geothermal, solar and wind.[11] Nuclear energy delivers about 6.5 per cent.[11] While renewables and perhaps even nuclear will contribute an increasingly higher share of the world's energy needs, the bulk of energy services over the next 20 to 30 years will continue to be met by the conventional energy sources, that is, oil, natural gas and coal.

Oil and gas reserves are geographically concentrated in a few regions of the world, with about 60 per cent of global oil reserves and 40 per cent of natural gas reserves located in the Middle East.[12] With the import dependency of the largest energy consuming countries projected to rise over the medium term, their reliance on hydrocarbon supplies from a limited number of producing regions, some of which are likely to continue to be subject to political and social unrest, will corresponding grow. However, the same is true for the large hydrocarbon exporting countries who will increasingly become ever more dependent on the oil and gas related revenues coming from the net energy importing countries. This growing interdependence, if properly managed, could prove mutually beneficial to both groups of countries.

On the other hand, if not properly managed by the international community, the global energy outlook might become even more uncertain and insecure than it is today. The key factors worsening global energy security problems today include the following:

- rapid growth in energy consumption, not only in developed but particularly in developing countries;

- slow pace of development of new reserves, lagging well behind the growth of hydrocarbon consumption;
- continuing low level of oil production in Iraq;
- volatility of world oil prices and their impact on other energy prices;
- insufficient pace of development of alternative energy sources;
- growing competition for access to energy resources;
- hydrocarbon related transit disputes; and
- environmental consequences of energy production, transport and consumption.

International energy security is traditionally considered largely from the standpoint of energy consumers or net energy importing countries. However, producers and exporters of energy resources face similar risks. Therefore, energy security needs to be addressed in a way that takes account of the interests of both groups. Unfortunately, the search to balance the interests of both groups is often difficult and complex. Nonetheless, these issues need to be addressed and resolved through multilateral and bilateral relations. Where possible, existing forums should be used to address energy security and related issues. For example, the United Nations could be more actively involved in addressing issues related to energy, since energy security can be undermined by a whole range of factors, such as political instability, conflicts, terrorism, poverty, social unrest and economic underdevelopment, all areas of particular relevance to United Nations.

Among the countries of Eastern Europe, the Caucasus and Central Asia, the Russian Federation, Kazakhstan, Azerbaijan, and Turkmenistan are major producers and exporters of hydrocarbons. The Russian Federation is of particular importance to the global energy market. in terms of global energy supply. With the largest estimated proved reserves, the Russian Federation is the world's largest producer of natural gas as well as its largest exporter. It is also the world's second largest producer and exporter of crude oil, with the seventh largest oil reserves. It is also well endowed with coal reserves (second to the United States), uranium and hydro resources.

In addition to being a major and secure producer and exporter of hydrocarbons, the Russian Federation also contributes to global energy security in a variety of other ways. It is continuing to develop its nuclear industry, including working on the development of fast reactors. It is striving to adjust its contractual arrangements with consuming and transit hydrocarbon countries so as to make these more market based and, thereby, contribute to enhancing energy security for all. Furthermore, Russia is actively promoting international cooperation in areas, such as, energy conservation and efficiency, renewable energy and research and development in new energy technologies.

Notwithstanding the above, one of Russia's main strategic interests is to diminish its economic dependency on the energy sector by diminishing the current large share of energy production and exports in total gross domestic product (GDP) through the greater development of competitive industries in other sectors of the economy based on high and mid-level technologies.

<u>Basic Trends in Global Energy Development</u>

According to current forecasts, world energy demand will grow by more than 50 per cent over the next 25 years.[13] Projections to 2030 show that coal, natural gas and petroleum are likely to continue to supply the bulk of energy services demanded, not much different than their current 87 per cent share of total global energy demand.[13] Therefore, over the medium term, these three energy sources are likely to remain the most common globally traded energy resources.

The rate of growth in supply of renewables and new alternative energy sources should exceed that of the conventional sources of energy but their share in total global primary energy demand is not likely to change much over the medium term because of their low starting base. The International Energy Agency expects that they could increase their share of total primary energy demand by about 1 percentage point by 2030.[14] The outlook for nuclear energy, which currently provides about 6.5 per cent of total global energy needs, is more problematic. Its future development is clouded with uncertainty because of concerns regarding the potential, even if remote, of nuclear accidents and the lack of adequate methods for disposal of nuclear wastes.

The multi-purpose character of oil, its availability and its ease of transport have made oil a unique type of raw material. It satisfies a third of the world's energy needs, it is at the centre of every nation's energy systems and the price of crude oil influences most of the other energy prices to various degrees. Its use is projected to grow by 40 per cent by 2030 and the required investment needs by the oil industry to meet this growth is estimated to be about $US 4.3 trillion (in 2005 dollars).[13]

Oil production is expected to rise most in the Middle East, Africa and Latin America. For example, the share of Middle East and North African countries in total oil production is projected to rise from about 35 per cent to about 44 per cent by 2030.[15] At the same time, the share of production of countries, such as, the Republic of Iraq, the State of Kuwait, and the Socialist People's Libyan Arab Jamahiriya in regional production is also expected to rise.

The oil import dependency of the European Union countries, the United States, China, India and other major oil importing countries is bound to rise over the medium term. The International Energy Agency has analyzed a number of scenarios in its world energy outlooks and all point to an invariable increase in oil import dependency for the major oil consuming and importing countries. The major difference is that the rise in imports and import dependency would be less if the major consuming countries introduced measures to reduce their vulnerability to energy supply risks and/or implemented measures to introduce cleaner energy sources and new technologies.[15]

There are several noteworthy features of the geographic location of oil reserves. First, oil resources are highly concentrated in relatively few regions of the world. Second, economically developed countries possess less than 20 per cent of the world's oil reserves. More than 60 per cent of global oil reserves are concentrated in the countries surrounding the Persian Gulf. As a result, the oil import dependence of the largest energy consuming nations will grow over time and the dependence of oil exporters on export revenues will likewise continue to rise. Consequently, risks related to energy transit and security can be expected to intensify. Following its extraction, more than half of the oil produced crosses international borders and often several of them before reaching consumers. This is especially significant for the European Union. As indicated earlier, 50 per cent of energy needs in the European Union are covered by imports today, a figure that is likely to rise to 70 per cent in 20 years.[16]

Over the last 10 to 15 years, natural gas has become the fuel of choice, with its market share rising significantly, for a variety of reasons. According to the IEA, the rate of growth of natural gas demand will exceed that of the other fossil fuels, primarily due to strong demand from the power generation sector. By 2030, natural gas will be competing with coal to become the world's second most widely used primary energy source. With the expanded production and use of liquefied natural gas, increasingly the natural gas market will also become more global than regional in character.

At present, the world oil market is still fairly segmented. This means that refineries located in certain regions depend on oil coming from specific oil fields. Oil produced in different geological conditions and, accordingly, in different regions and countries, differs by composition and quality (content of paraffin, sulphur, mercaptan, heavy bitumen, light, and volatile fractions) as well as by physical and technological properties. Many refineries have a rather limited capacity to adapt to different varieties of oil. As a result, if a crisis were to arise in a particular oil producing country or province, this segmentation could lead to local deficiencies of crude and consequently to local shortages of petroleum products.

In some regions of the world, expansions in oil refinery capacity are substantially lagging behind the growth in consumption of petroleum products. This applies particularly to the United States and some other developed countries where authorizations for new large energy infrastructure projects are subject to lengthy review processes and are frequently opposed by groups in society.

The review process by regional and local authorities for the construction of new refineries in the United States and the European Union can take several years to meet land use planning specifications, and environmental, health and safety regulatory requirements. Various environmental restrictions introduced in the United States and the EU over the last 20 to 25 years have contributed to lengthening the review process and, thereby, generating greater uncertainty and additional costs to investors. In addition, different states in the United States set different fuel quality standards. This can potentially restrict the room for manoeuvre during times of shortages involving the production of gasoline and other types of fuel.

In the United States, this problem arose in the third quarter of 2005 as a result of Hurricane Katrina, which significantly disrupted the petroleum-product segment for several months. Nine refineries with an overall consumption of 1.5 million barrels of crude oil per day (b/d) suffered significant damage. This contributed to a sharp rise in gasoline and other petroleum product prices. In September 2005, the United States faced an almost two-fold oil price rise.

The United States does not possess state-owned strategic stocks of gasoline or diesel fuel that could be used to stabilize consumer prices during periods of shortage. Consequently, a disruption in gasoline or diesel fuel production, resulting in a significant shortage, could potentially contribute to a strong upward spiral in oil product prices. The strategic overall stock of liquid fuel in the United States, which is about 700 million barrels of oil with Congressional authorization for its eventual increase to one billion barrels, is judged by some analysts to be insufficient when compared to the current US consumption of 21 million b/d.

Wary of crises compromising its energy import-dependent economy, Japan has built up substantial strategic stocks of both crude oil and petroleum products, including gasoline. According to some estimates, it is equivalent to 160 to 180 days of Japan's current consumption rate. The prevailing view in Japan is that these stocks should only be used in extreme situations, which, according to an assessment by the Japanese Government, did not exist following Hurricane Katrina.

At present, two main types of economic entities dominate the world hydrocarbon sector, private oil and gas companies and national oil and gas producing companies owned by the main oil-producing states. Both are actively engaged in the development of new oilfields and major large-scale projects. The national oil and gas companies are strategically important since they control access to some of the most promising oil and gas reserves. Energy Intelligence, an international research organization, estimates that these companies control access to 57 per cent of all proven oil reserves and over 45 per cent of all hydrocarbon production.[17] According to James R. Schlessinger, former Secretary of Energy of the United States, practically all strategic oil reserves are concentrated in national state-run companies.[18]

Various assessments by Western analysts indicate that about two thirds and, most probably, more of all hydrocarbon resources are inaccessible to private-sector international oil and gas companies. In some countries, such as the United Mexican States (Mexico) and the Kingdom of Saudi Arabia, all reserves, for all intense and purposes, are fully closed to foreign investment.

Over the last 15 to 20 years, global energy markets have increasingly been opened up and liberalized, with government intervention in the energy sector receding. But as noted in a number of earlier sections, this trend may now be reversing and the pendulum swinging back to more state intervention (i.e., greater statism). This has occurred in the Russian Federation where there has been some roll back of post-1990 measures, particularly in 2004 and 2005, with respect to privatization and foreign ownership. The state has consolidation its ownership and control over the oil and gas sector in part to correct some recent past excesses as well as to ensure that the benefits to Russia from oil and gas development are maximized, in view of the importance of the hydrocarbon sector to the Russian economy. Energy accounts for about 55 per cent of total export revenues and oil and gas activities provide up to 50 per cent of federal budget revenues.

General Risk Factors of Energy Security

Energy security has today become a very pertinent issue for high-level policy discussions.[19] The main reasons for this include the following:

- **The growth in energy consumption** especially of hydrocarbons has outstripped the rate of growth of production capacity, thus leading to a tightening of energy markets. This has manifested itself through a reduction in the availability of standby producing capacities, particularly in the oil sector. This situation is likely to persist for some time. After the energy crisis of 1973-1974, it took about 10 years to transform the oil market from a supply-constrained to a demand-constrained market. Over that period, high oil prices boosted the introduction of energy saving technologies on the one hand and encouraged investment in the petroleum industry and in the development of alternative energy sources, such as nuclear, on the other hand.

- **The pace of discovering and developing new deposits** is lagging behind the growth in consumption of hydrocarbons, which is mainly linked to the deficiency of energy investments. At their Gleneagles Summit in July 2005, the G8 leaders stressed the need for substantial current, middle-term and long-term investments in exploration, energy production and development of energy infrastructure.[20]

- **The continued low level of oil production in Iraq** which has exacerbated the already tight oil market. In early 2006, oil production was still lower than before the military conflict of 2003. Furthermore, there continues to be a lack of sufficient new investment in the Iraqi petroleum industry due to the challenging conditions. The penetration of private western companies in Iraq is complicated not only by the high violence level and direct danger to the life of employees but also by the general political uncertainty in Iraq.

- **The volatility of world oil prices** combined with their general higher level has raised concerns about the future availability of hydrocarbon resources and the implications for economic growth. The G8 statement on the "Global Economy and Oil" issued at the Gleneagles Summit stated that high and volatile oil prices were a "challenge" to world economic growth. It also noted that the strong global growth boosted energy demand and, together with capacity constraints and supply uncertainties, contributed to high and volatile oil prices."[20]

- **Competition for access to energy resources** both among companies and among different states has intensified, straining overall energy relations. The bargaining power and 'strategic landscape' of energy security has changed over the past decade in favour of hydrocarbon producers over energy consuming and importing countries. It has also changed in part due to the significant rise in energy demand of China, India and other developing countries. While pursuing their national interests and trying to preserve their high rate of economic growth, these countries have, at times, aggressively pursued additional supplies of hydrocarbons by offering much better terms to hydrocarbon-rich countries for developing their reserves compared to what western transnational companies were able to offer.[21] In addition, China has strengthened its relations with OPEC. In December 2005, China and OPEC had official talks in Beijing in order "…to establish a balanced, pragmatic framework for cooperation, and to develop an ongoing exchange of views at all levels on energy issues of common interest, in particular security of supply and demand, in order to enhance market stability."[22]

- **The problems of selecting transit routes for oil and gas pipelines** and the discord between countries over route selection. To appreciate the drivers behind route selection, one needs to take into account not only their direct economic justification and the degree of political stability in the countries where they are laid, but also the degree to which they match the general geopolitical interests of states for whom the projects have relevance. Contemporary transit problems were vividly illustrated, for instance, in the process of building the Baku–Tbilisi–Ceyhan oil pipeline. In addition, while choosing an itinerary for the oil pipeline from Russia to China in 2000-2001, Chinese authorities rejected the proposal of laying the pipeline through the territory of Mongolia, although this was the most expedient route from an economic standpoint.

- **Disagreements arising within the "triangle" of producer-transit-consumer countries,** which normally should have a mutuality of interest and interdependence. Two recent examples are the Russia-Ukraine disagreement in 2006 over natural gas and the Russia-Belarus disagreement over oil in 2007 regarding the terms of supply and transit for these commodities. The ensuing short-run disruptions in Russian deliveries of natural gas and oil to countries in Central and Western Europe generated serious concern in downstream consumer countries.

- **Territorial disputes over areas endowed with significant energy deposits.** There are a number of such disputes. For example, the dispute between China and Japan over the Senkaku (Diaoyutai) Islands is a case in point. The dispute between Iraq and Iran with respect to the regime governing the use of the border waterway Shatt El Arab is another example. This latter dispute has lasted for several decades and been the source of conflict. Control of oil has also been behind the scenes of other conflicts, including the Ethiopian-Somalian and Vietnamese-Chinese conflicts in 1979. In Nigeria, the Republic of Angola and the Republic of Indonesia, separatists have actively conducted military operations, but largely in the oil-rich provinces. Indeed, historical precedents might suggest that many political and military strategies are, in many respects, designed to guarantee the security of hydrocarbon access, secure transportation or to benefit local populations from oil and natural gas revenues.

- **Environmental factors over the last 25 years have increasingly impacted on energy decisions with implications for energy security.** Firstly, "energy ecology" involves the problem of harmful effluents from energy production. Secondly, there is the problem of accidents while transporting the energy resources, and in particular if damage occurs to pipelines. Thirdly, there is the issue of damage to the environment arising from the processing of energy resources, especially in refineries, and their ultimate consumption. Fourthly, the energy industry tends to require significant amounts of water and in fact consumes some two thirds of the clean water used by all industries.

Political Risks

Generally speaking, a range of political risks have existed for many decades:[19]

- **internal political instability** causing unrest, including civil wars and inter-ethnic conflicts, which reduce the production or transportation of oil and gas and thereby curtail output;

- **interstate wars** and military conflicts in the regions of oil and gas production and transportation; and

- **terrorist acts** against oil and gas production areas, and critical infrastructure.

There are increasing concerns also regarding transit bottlenecks, not only through the Straits of Ormuz, the Bosphorus and the Dardanelles, but also through the Malacca Straits where the navigable part used by supertankers is only 500 meters wide and 10 meters deep.

The greatest danger to the stability of the world economy would arguably come from two or more simultaneous serious energy incidences. In order to mitigate the potential repercussions of such risks, it would be prudent to establish mechanisms for international crisis management in advance. Such a mechanism could seek to address the legitimate interests of all parties concerned in an acceptable manner. The International Energy Agency has significant experience in this respect, but its membership is limited to highly developed countries that are net hydrocarbon importers. As a result, energy exporters often perceive the IEA as an organization reflecting the interests of energy importers. It might therefore be better to create such a mechanism under the auspices of the United Nations as an institution that could address the interests of both the consumers and producers of energy resources.

International energy related political risks, however, are not confined to the production and transportation of hydrocarbons. They are also related to potential acts of terror targeted at power facilities and their related critical infrastructure. Many countries have recently taken substantial measures to increase the protection of such facilities in an attempt to contain the potential impact of terrorism. But the fact that terrorists are constantly learning and improving their methods of attack is of particular concern.

Minimizing political risks and the development of respective risk management systems may be considered as one of the most important ways of maintaining global energy security. Risk management is important both economically and politically. In the economic context, it can reduce the costs of financing investments thereby ultimately enhancing the performance of a nation's economy. Politically speaking, risk management can help to prevent or reduce social and political unrest inside countries as well as prevent or deter conflicts between them.

International energy security has traditionally been considered primarily a concern of energy importers who wish to maintain a stable supply of hydrocarbons at reasonable prices. In order to accomplish this, energy producers and exporters have generally maintained significant standby facilities which can be used during an oil crisis to help compensate for reduced deliveries from one or more of the exporting countries by expanding the oil production and deliveries from other countries or regions. Likewise, strategic and commercial stocks have been used to mitigate the effects of supply shortages due to major energy crises.

From the producer's point of view, ensuring the security of energy supplies in this way carries with it substantial risks. These energy security risks include the:

- **cyclical nature of the world economy** with its rising and falling demand for energy resources;

- **higher investments** needed to expand energy production and transportation systems in order to provide for fluctuating demand and unforeseen contingencies; and the

- **declared intentions** of long-standing consumers to turn to alternative energy sources.

In the past, embargoes on the delivery of energy resources by some producers have also contributed to destabilizing energy markets.

While consumers depend on energy exporters, producers are also highly dependent on consumers. The majority of state revenues in OPEC and other oil producing countries are derived from oil and natural gas exports. Producers not only have to contend with the risks cited above, but they are also affected by the fluctuations in value of the United States dollar, the main currency used to trade oil. Given the complexity of these risks and the different but commensurate risks and the mutuality of their interests, energy security burden sharing should be based on an equitable sharing of risks between producers and consumers taking into account their mutual economic interdependence.

But today, the question of burden sharing between producers and consumers is further complicated by the need to take into consideration a host of other factors, such as, political, social and security factors. Some projects, which in the past would have been evaluated solely on economic grounds or on a "cost-efficiency" basis, now have to be evaluated against a wider range of criteria. For example, military and political security considerations can create additional uncertainty, risk and attendant costs for ensuring energy supplies at reasonable prices to consumers.

Reconciling the Global Interests of Energy Importers and Exporters

Energy importers are naturally interested in modest or low prices for the energy they import. Despite this, some hydrocarbon importers such as the United States are also producers themselves who cannot afford to see oil prices fall below a certain level. If this were to happen, its own oil production would become uncompetitive with cheaper oil from the Persian Gulf and the Bolivarian Republic of Venezuela.

In contrast to importers, the interest of net energy exporters is to have high oil prices over the long-term and stability in consumption. However, if hydrocarbon prices are too high over a prolonged period of time this increases the likelihood that alternative energy sources become economically viable and deployed in a significant way. Furthermore, some net exporters of hydrocarbons might prefer not to have excessively high prices since these could in the longer-term negatively impact their own economy through inflationary pressures affecting other sectors of their economy (though it should be noted that there are ways of handling high oil and gas revenues so that these do not create domestic inflationary problems). Very high prices also increase the share of energy in industrial output and in the national economy making the economy more vulnerable to a downturn in hydrocarbon prices. Furthermore, high prices promote the accelerated exploitation of energy reserves, which ultimately could reduce the amount of recoverable reserves, diminishing the long-term production and export volume of hydrocarbons.

The concept of an equitable oil price and equitable prices for other energy commodities is very difficult and complex to ascertain. What may be equitable to some may not be so to others. Despite this, some producer organizations/associations and producer-consumer groups have occasionally attempted to define prices or a price range that they judge to be "fair". OPEC countries, for instance, have periodically tried to support prices within a given range through quotas for producing oil delivered to the world market. For example, in 2005, OPEC decided to keep oil prices over the 2006-2007 period within a range of US$ 45 to US$ 55 per barrel. In the end, because of market conditions, prices were allowed to rise well above this range.

Energy importers are interested in diversifying the sources of their energy imports while net energy exporters are interested in diversifying their markets. These preoccupations also concern the Russian Federation. Diversifying markets and sources of energy supplies depends greatly on the adequacy of energy transport infrastructure. This includes pipeline networks, railway transport, terminals, liquefied natural gas facilities and refineries. Developing and maintaining this infrastructure requires enormous investments. For the private sector, such large long-term investments carry with them considerable uncertainties and risks.

At the same time, the interests of net energy exporters and net energy importers have much in common, especially in maintaining the efficient flow and trade in energy as well as supporting the stable growth of the world economy as a whole. This co-dependence or interdependence among net exporters and net importers of hydrocarbons perhaps deserves greater attention given the strained relations sometimes between energy importers and exporters.

As noted earlier, the search for the "balance of interests" between consumers and producers is a complex multidimensional and difficult exercise. The G8 Summit process and other international forums have played an increasingly important role in this respect in recent years. However, the problem of enhancing the mechanisms for resolving international energy security problems continues to elude the international community.

At the G8 Summit held in Gleneagles, participants agreed that energy resources were fundamental for economic stability and development. The recommendations from this Summit meeting called on the oil producing countries to take all possible measures to improve their investment climate. Oil producing countries were urged to ensure open markets with transparent business practices and stable regulatory frameworks for investment in the oil and gas sectors, including increased opportunities for foreign investment. The importance of the dialogue between the oil producing countries and oil consuming countries within the framework of the International Energy Forum (IEF) was underlined. The declaration also stated the need to encourage the expansion of refinery capacity.[20]

But despite the G8 declaration and the numerous other bilateral and multilateral discussions on energy issues, there continues to be deep-seated divergent views among countries on energy. For instance, Russian policymakers viewed negatively and with suspicion the decision by the United States and Western European interests to bypass Russian infrastructure by building the Baku–Tbilisi–Ceyhan Pipeline to move oil from the Caspian Sea region through Turkey and beyond. Similarly, United States policymakers reacted rather negatively to the proposed Iran–Pakistan–India Pipeline Project. Furthermore, when the China National Offshore Oil Corporation (CNOOC) expressed its intention to buy the Californian oil company Unocal, the United States Congress nearly adopted a law banning the sale of petroleum assets to Chinese companies for national security reasons.

At the same time, western oil companies have become quite active in China. Jointly with Chinese companies, Shell owns the largest multi-billion petrochemical company in Shenzhen located in the south of the country. While Shell's main activity in China today is related to oil refining and the sale of fuel and petrochemical products, it is planning to become more actively involved in prospecting and the production of hydrocarbons. The company is actively prospecting for oil and gas both on land and in shallow waters offshore China. ChevronTexaco and CNOOC are also actively cooperating in China. Both companies are partners in one of the largest oil-and-gas projects in the world: a twenty-five year contract involving the delivery of liquefied natural gas from Australia to China at an estimated value of US$ 35 billion.

Russia's Global Energy Role

The Russian Federation has 26 per cent of the world's proved natural gas reserves, 7 per cent of the world's oil reserves and 17 per cent of coal reserves.[12] It also has substantive reserves of uranium and significant hydro resources. The Russian Federation is the world's largest exporter of natural gas and the second largest oil exporter. After the United States (which consumes about 20 per cent of the world's primary energy) and China (about 15 per cent), Russia (about 7 per cent) is the world's third largest consumer of primary energy resources. The energy sector is one of the most stable and expanding sectors of the Russian economy.

The fuel and energy sector of the Russian Federation is heavily oriented to exports. In 2005, Russia exported over 50 per cent of its crude oil output, more than 40 per cent of its petroleum products, about 30 per cent of its natural gas and 25 per cent of its coal production. In 2004, the hard currency income from fuel and energy products amounted to more than half of the country's export earnings. The main sources of this hard currency came from exports of oil, natural gas, and petroleum products as shown in Table 1.5.1.[23]

Table 1.5.1 Russian Federation Energy Exports in 2004

Energy Type	Million Tonnes (Mt)	Export to Production Ratio (per cent)	Commodities Export US Dollars (billions)	Ratio of Energy Export to Total Russian Exports (per cent)
Oil	257.1	56.0	58.2	31.9
Petroleum products	82.0	42.1	19.1	10.5
Natural gas, billion cubic metres (bcm)	200.4	31.7	21.9	12.0
Coal	72.0	25.7	2.8	1.5
Electric power, billion kWh	17.7	1.9	0.5	0.3
Total			102.5	56.2

Sources: Rosstat and the Russian Customs Agency. A. S. Nekrasov, Economic Problems and Future Outlook of the Russia's Energy Sector. Presentation to the General Meeting of the Russian Academy of Sciences, 20 December 2005, p. 4.

The Russian Federation has resumed its position in the global oil market that it lost following the collapse of the Soviet Union. Over the last six years, crude oil production has risen from 6.2 to 9.8 million b/d in 2006.[12] However, the pace of oil production growth has recently slowed down significantly, increasing by 2.2 per cent over 2005.

A detailed breakdown by country showing oil deliveries from Russia and national oil consumption is presented in Table 1.5.2. It shows that Russian oil is the key component in the oil consumption of Eastern European countries, the Baltic States and the Republic of Finland. The Swiss Confederation (Switzerland) with 40 per cent of oil consumption originating from the Russian Federation, the Republic of Austria with 24 per cent, Italy with 23 per cent, the Federal Republic of Germany with 24 per cent and the Kingdom of the Netherlands with 21 per cent are also significantly dependent on imported Russian oil.

The growth of Russia's share in total world energy exports is partly due to the significant increase in exports of natural gas. These in turn have risen because of the rapidly improving technologies for the production, transportation and use of natural gas. These exports could continue to rise further over the near to medium term since, according to some assessments, Russia's proved natural gas reserves could double and reach 100 trillion cubic metres in the future. However, such assessments are uncertain until proven by expensive geological prospecting and then there is the question of developing these new reserves which increasingly will be in remote areas and subject to harsh environmental conditions.

Data from the Ministry of Industry and Energy show that in 2005 gas production in the Russian Federation reached 640.6 billion cubic metres (bcm), about 1 per cent higher than in 2004. At 547.3 bcm, Gazprom accounted for 85 per cent of total production. Natural gas exports of 152.5 bcm in 2005 were 8 percent higher than in 2004.

Table 1.5.2 Russian Federation Oil Exports

Country	Oil Deliveries from Russia (thousand tonnes)	Share of Russian Oil in National Oil Consumption (per cent)	Share of Russian Oil in Total Oil Imports (per cent)
Latvia	1866	100	100
Lithuania	8661	100	100
Slovakia	5551	98.8	99.2
Poland	17181	98.3	99.4
Ukraine	19091	90.6	100
Hungary	5273	79.7	100
Finland	7692	69.9	70.4
Czech Republic	4452	67.7	69.4
Kazakhstan	3153	53.3	100
Bulgaria	2411	46.1	46.2
Croatia	2116	43.8	54.3
Switzerland	1809	40	40
Romania	3997	29.8	62.8
Germany	26 395	24.2	24.8
Austria	2149	24.1	29.3
Italy	20 907	23.5	24.9
Netherlands	10 864	21.5	22.4
Sweden	4030	19.5	19.5
Israel	1667	17.4	17.4

continued .../

Table 1.5.2 Russian Federation Oil Exports *continued*

Country	Oil Deliveries from Russia (thousand tonnes)	Share of Russian Oil in National Oil Consumption (per cent)	Share of Russian Oil in Total Oil Imports (per cent)
Serbia and Montenegro	423	16.1	22.6
Turkey	3627	13.6	15.1
Morocco	838	13.2	13.2
Greece	1387	7	7
France	5531	6.4	6.5
Spain	3145	5.5	5.5
United Kingdom	3330	4.3	23.8
Egypt	1064	3.9	20.4
China	7365	3.2	10.6
Portugal	380	3	3
Belgium	751	2.1	2.1
Republic of Korea	1515	1.4	1.4
United States	7886	1	1.5
Norway	113	0.8	2.9
Japan	1442	0.7	0.7
India	135	0.1	0.2

Note: Data for 2003. Slight inaccuracies are possible due to the exclusion of oil deliveries to offshore areas where some oil traders exporting Russian oil are registered. Sources: BP Statistical Review of World Energy, 2005; data from Rosstat; "Gazprom"; IEA "Oil Information 2004"; IEA "Energy Statistics of Non-OECD Countries 2002-2003"; Customs Statistics of the Russian Federation, 2003. Data at slight variance with those contained in a similar table published in "Nezavisimaya Gazeta" on December 2005.

Gazprom is the country's largest gas producing company. It also owns all the main gas pipelines in Russia and markets a significant portion of the natural gas produced in the Caspian Sea region. The company's capitalization changed radically when trade in Gazprom shares was liberalized in December 2005 and opened up to foreign investors. As of January 2006, the company's capitalization exceeded US$ 200 billion and it ranked, on the basis of its capitalization, as the world's seventh largest company.

At present, Russian gas covers 26 per cent of gas consumption in Europe and about 40 per cent of European imports.[24] About half of all the gas consumed in the European Union is sourced from only three countries: the Russian Federation, the Kingdom of Norway and the People's Democratic Republic of Algeria.[25]

An overview of Russian natural gas exports is provided in Table 1.5.3. As with petroleum, Russian natural gas exports provide most of the needs of Eastern Europe, the Former Republics of the Soviet Union and also of Finland. In contrast to oil exports, however, Russian natural gas is in a much more favourable position in Europe due to the proximity of Russia to major European markets and fewer competing sources of supplies. For example, Greece, Turkey and Austria get more than half of their natural gas supplies from Russia. Germany at 43 per cent and France and Italy at around 30 per cent are also important markets for Russian natural gas.

Today, Gazprom and other gas producing companies in Russia largely produce and market pipeline gas. But to diversify its exports, Gazprom is planning to enter the LNG market in order to be able to access the American and Asian markets. Inter-regional trade is expected to grow significantly over the medium term with new liquefaction, shipping and regasification capacity being built. Gazprom has already acquired a majority interest in the Sakhalin II Project and is considering how to develop the Shtokman field in the Barents Sea.

Table 1.5.3 Russian Federation Natural Gas Exports

Country	Gas Deliveries from the Russian Federation (billion cubic metres)	Share of the Russian Federation in Aggregate Gas Consumption (per cent)	Share of the Russian Federation in Aggregate Gas Import (per cent)
Moldova*	2.7	245.5	100.0
Serbia and Montenegro	2.3	100.0	100.0
Estonia	0.9	100.0	100.0
Bulgaria	3.1	99.6	100.0
Finland	4.3	99.2	100.0
Latvia	1.5	93,8	93.8
Lithuania	2.9	93.2	93.5
Greece	2.2	90.0	80.0
Slovakia	5.8	85.6	79.5
Czech Republic	6.8	76.5	69.4
Hungary	9.3	71.5	84.9
Turkey	14.5	65.3	65.3
Austria	6.0	63.5	76.9
Belarus	10.2	55.3	51.5
Ukraine (2004)	34.3	48.5	50.4
Poland	6.3	47.6	69.2
Germany	37.3	43.4	40.6
France	13.3	29.8	29.8
Italy	21.6	29.5	35.2
Romania	4.1	21.8	69.5
Switzerland	0.3	10.0	10.3
Netherlands	2.7	6.2	19.9
Kazakhstan	0.8	5.3	n/a
Belgium	0.2	1.2	1.0

Sources: Rosstat and the Russian Customs Agency statistics. A. S. Nekrasov. Economic Problems and Future Outlook of the Russia's Energy Sector. Presentation to the General Meeting of the Russian Academy of Sciences, 20 December 2005

Russia's Oil and Gas Resource Base

In 2003, the Government of the Russian Federation approved "The Energy Strategy of Russia Until 2020." Given a price of US$ 22-26 per barrel and favourable market conditions, oil production is projected in the strategy to reach 490 Mt by 2010 and 520 Mt by 2020 compared to 470 in 2005.

Natural gas production is projected to reach 665 bcm in 2010 and 730 bcm in 2020 compared to 641 bcm in 2005.[26] There is, however, some doubt about whether the oil projections are achievable given the current slowdown in the rate of growth of oil production. On the other hand, natural gas resources are much more plentiful and should allow Russia to meet the projected volumes and perhaps even surpass them.

The Russian Federation possesses several resource basins for oil and gas production. There are the oil and gas bearing regions of the European part of the country: Volga-Urals, North Caucasian, and Timano-Pechora Province, as well as the Russian sector of the Caspian Sea region. Over the coming decades, however, oil and gas production is expected to decline both in the Volga-Urals and, in particular, in the North Caucasian oil-gas-bearing Provinces that have been providing Russia's economy with oil and gas for most of the twentieth century. There have been some attempts to expand oil and gas production in Northwest Russia and a number of large deposits are to be developed in the Timano-Pechora Province. The total production of oil in the European regions of the Russian Federation could reach 120 Mt.[26]

The Barents Sea shelf offers an exceptionally rich hydrocarbon potential for Russia. The giant Shtokman gas condensate field could provide natural gas production of 75-90 bcm per year in the future. Large oilfields have also been discovered in the Pechora Sea.[26] The Russian sector of the Caspian Sea – along with that of Kazakhstan – is especially rich in energy resources. This can be seen from the extensive deposits at Rakushechnoye recently discovered by Lukoil, which contain some 700 Mt of oil and over 1 tcm of natural gas. In addition, it is a promising region for new production being located in the eastern portion of the European part of Russia with a highly developed transport infrastructure.
The second oil and gas producing region of the country is located in Western Siberian and this Province will continue to have the largest output in the future until perhaps 2045 – 2050. There are extensive undeveloped oil and gas resources concentrated on the Yamal Peninsula. A number of large deposits are to be developed in the Nadym-Purskoye Interfluve in the north of the Province (Yamal-Nenets Autonomous Area) and in its central regions (Khanty-Mansi Autonomous Area). Oil and gas production is also growing rapidly in the Tomsk Region. Crude extraction has recently started in the south of the Province, notably in the Tyumen, Omsk, and Novosibirsk Regions. In 2003, this Province produced more than 300 Mt of oil and 570 bcm of natural gas. By 2020, the production in the Province will amount to at least 315 Mt of oil and gas production may reach 680 bcm.[26]

The third base of the oil and gas industry in Russia to be developed within the next 10 to 15 years is in Eastern Siberia, including the Republic of Sakha (Yakutia). The Siberian Platform is geologically similar to the African Platform, which is located in Northern Africa to the south of the Sahara. Harsh conditions have so far negatively affected the development of the oil and gas potential of the Siberian Platform. Nevertheless, in the 1980s, more than 30 oil and gas deposits of different sizes were discovered and explored, including such gas giants as Kovyktinskoye and Chayandinskoye, as well as large oil and gas bearing fields, including Yurubcheno-Takhomskoye, Verkhnechonskoye and Talakanskoye. Taking account of the degree of exploration undertaken to date and the required infrastructure, oil production in Eastern Siberia could reach 10 Mt by 2010 and up to 60 Mt by 2020. Natural gas production could rise to 30 bcm by 2010 and 115 bcm by 2020. Crude oil extraction is planned to begin from the Yurubcheno-Takhomskoye, Verkhnechonskoye and Talakanskoye deposits and similarly gas production could commence in the near future at the Kovyktinskoye deposit. However, it needs to be borne in mind that all these deposits require additional prospecting and new transport infrastructure which, in turn, will require very large investments.

The oldest sedimentary rocks that form the gas deposits in Eastern Siberia and the Republic of Sakha contain helium in industrially significant quantities. Together with the United States, Eastern Siberia is a strategically important region with regard to natural gas deposits containing helium resources.

The fourth largest base of Russia's oil and gas industry is taking shape in the Far East on the Sakhalin shelf. Oil production in this region may reach 20 Mt per year and the production of natural gas could rise to 25 bcm per year. These resources are being developed mainly for export to the United States and countries of the Asian Pacific region.

The Russian gas supply system was initially developed to be an integral system for the enormous Eurasian space and also, to a considerable extent, to provide exports to foreign markets in Central, Eastern and Western Europe. In particular, Russia's system has many underground storage facilities, which serve to increase the stability of the system as a whole. This aspect also allows Russia to optimize its delivery of gas resources. In recent years, Gazprom has made significant efforts to develop and improve this system, which was inherited from the Soviet Union times. Gazprom has about 140 thousand kilometres (km) of key gas pipelines.

The Russian Federation has a rich tradition in geological research and prospecting for energy resources. State-of-the-art technologies can be assimilated relatively quickly, including secondary and tertiary recovery technologies, which can contribute to expand existing reserves and to increase the yield of operating wells. Russia has significant potential for enhanced recovery of hydrocarbon, notably in the Western Siberian oilfields. During the period of economic transition to a market economy, Russia accumulated considerable experience in international energy cooperation with foreign partners. This cooperation, particularly on advanced technologies with western companies, allowed Russia to noticeably increase the level of oil extraction from certain deposits.

To maintain and increase fuel and energy supplies, Russia will have to invest heavily in upstream and downstream infrastructure. According to some estimates, the minimum level of investment in the oil sector that will be required by 2020 is likely to be in the order of US$ 200-210 billion. Securing future oil and gas extraction in the required volumes will require an additional US$ 155-160 billion.[27] Clearly, the investments required are very large. Given current conditions, financing will have to come not only from domestic sources but also from foreign sources, including not only from western countries but also from eastern countries, such as China and India. Fortunately, the current high prices for hydrocarbons, if sustained in the future, can facilitate the flow of funds for the development.

Role of the Other Hydrocarbon Exporters in the CIS

The largest hydrocarbon reserves in the other countries of the Commonwealth of Independent States (CIS) are located mainly in the Caspian Sea region. According to the BP Statistical Review of World Energy (2006), Kazakhstan has reserves of 40 billion barrels of oil and 3 trillion cubic meters of natural gas, Azerbaijan has 7 billion barrels of oil and 1.4 trillion cubic meters of natural gas while Turkmenistan has 0.5 billion barrels of oil and about 3 trillion cubic meters of natural gas. However, estimates of proved hydrocarbon reserves for countries in the Caspian Sea region vary widely. For example, according to one Government source, Kazakhstan has reserves of some 51 to 60 billion barrels of crude oil, Azerbaijan 27 billion barrels and Turkmenistan about 16.1 billion barrels.[28]

Azerbaijan

The Republic of Azerbaijan has 0.6 per cent of the world's proved oil reserves. These reserves would be sufficient to maintain the current level of production for about 30 years (reserve/production ratio). However, the full potential of Azerbaijani deposits has not yet been fully evaluated. The Azerbaijani sector of the Caspian Sea has a relatively complex sub-surface geology, which, with further prospecting and new technology, could yield significantly higher volumes of oil.

Azerbaijan has a rich history of hydrocarbon production and a highly developed energy sector. The government has concluded more than 26 hydrocarbon production-sharing agreements with foreign partners. The Baku-Tbilisi-Ceyhan oil export pipeline came into service in May 2006. The Baku–Tbilisi–Ersurum gas pipeline is under construction. Both of these pipelines will contribute to diversifying the transit routes for export of hydrocarbons from the region.

Kazakhstan

During the 1990s, crude oil production in the Republic of Kazakhstan rose sharply. Production continus to rise as new oilfields are put on stream, new wells drilled, and old ones repaired. In addition, the application of enhanced petroleum extraction technologies is helping to increase oil recovery. Kazakhstan's oil and gas industry is currently developing some 80 hydrocarbon deposits.

By the end of this decade, Kazakhstan could be producing over 90 Mt of crude oil per year,[28] compared with 66 Mt in 2006. With only a small fraction consumed domestically, the remainder will be exported through three main pipeline systems: the Caspian Pipeline Consortium through Russia which will be expanded to transport 67 Mt per year; an oil pipeline to China commissioned in 2005 with capacity of up to 10 Mt; and the newly constructed Baku-Tbilisi-Ceyhan pipeline to which Kazakhstan will eventually be connected. For this third export route, crude oil will initially be delivered from the port of Aktau by tankers to Baku. Subsequently, a pipeline along the Caspian Sea bottom will be constructed to provide at least an additional 20 Mt of oil export capacity. The three pipeline systems should be able to handle Kazakhstan's total exportable volume of oil, which, according to current plans, is projected to be 120 Mt out of a total production of 150 Mt by 2015.

Kazakhstan almost quadrupled its production of natural gas over the last ten years to 23 bcm per year in 2006. By 2015, Kazakhstan expects another sharp expansion to 80 bcm per year.[28] While production is expected to rise significantly, domestic natural gas demand is not expected to appreciably increase thus leaving increasing amounts of natural gas available for export.

At present, 45 deposits, containing over 80 per cent of recoverable gas resources, are being developed. In 2004, 9 bcm or almost 44 per cent of total gas production originated from associated petroleum gases. Given the high proportion of gas in Kazakhstan's oil deposits, the output of associated gases will rise in the future in direct proportion to increasing oil production. The lack of adequate gas recovery systems, however, means that currently almost a third of the associated gas is flared or sometimes used to meet local oil production needs. Hence, with better recovery systems, more gas could be made available for export.

Significant gas transit systems criss-cross Kazakhstan. They are an important means of supplying gas to the entire territory of the former USSR. These transit pipelines connect the gas producing areas of Central Asia with consumers of the European part of Russia, the Trans-Caucasus, and, via Russia, with Ukraine and Europe. At present, the gas transit system 'Central Asia-Center' is being upgraded in order to increase its throughput from the present 54.6 bcm to 80 bcm. It will then be subsequently expanded to 100 bcm. Improvements to the system will be done in seven stages over nine years at a total estimated cost of about US$ 2 billion.

Turkmenistan

Turkmenistan is an important producer and exporter of natural gas but is currently not an important producer of oil. At present, Ukraine, Russia and the Islamic Republic of Iran import natural gas from Turkmenistan. Naftogaz Ukrainy is currently the largest importer of Turkmen gas. The Islamic Republic of Iran started importing natural gas from Turkmenistan in December 1997 when the gas pipeline from Korpeje (Western Turkmenistan) to Kord-Kuy (Northern Iran), with a capacity of 13 bcm of gas per year, was put into service. The volume of Turkmen gas deliveries to the Islamic Republic of Iran has gradually expanded from less than 2 bcm in 1998 to about 6 bcm in 2006.

The intergovernmental agreement with Russia on cooperation in the gas sector signed in April 2003 provides for the delivery of 1.8 tcm of natural gas to Gazprom until to 2028. In 2004, Russia imported 4.8 bcm of gas from Turkmenistan. According to the agreement, deliveries of Turkmen gas to Gazprom are expected to rise significantly. Deliveries in 2007 are expected to reach 50 bcm and in the future could rise to 80 to 90 bcm. At present, Turkmenistan is actively promoting a gas pipeline project to transit Turkmenistan, the Islamic Republic of Afghanistan and the Islamic Republic of Pakistan with a possible extension to India.

Since 2005, the Government of Turkmenistan has taken a number of steps to promote the future export of natural gas to China. The Chinese National Petroleum Corporation (CNPC) has been given the right to prospect some of the most promising geological structures on the right bank of the Amu Darya River. On the basis of these exploration results and a feasibility study, CNPC would decide whether or not to proceed with development of these deposits. The expectation is that, if the company were to proceed with development, it would be on the basis of a production sharing agreement and a commitment linking development of this promising area with an export pipeline to China.

While Turkmenistan's oil sector is relatively modest by world standards, there is the potential for oil production and exports to rise appreciably in the future. For example, the hydrocarbon potential in the Turkmen sector of the Caspian Sea is estimated at 11 billion tonnes of crude oil and 5.5 trillion cubic metres (tcm) of gas concentrated at depths of 2,000 to 7,000 metres. These estimates are based on a seismic survey conducted by WesternGeco, a United States company, under contract to the Government of Turkmenistan. Therefore, Turkmenistan could potentially also emerge as an important producer and exporter of oil as it will certainly be the case for natural gas.

Uzbekistan

The Republic of Uzbekistan's energy sector is based primarily on its commercially viable natural gas reserves. These are estimated at about 1.6 –1.9 tcm depending on source. In 2006, natural gas production was about 55 bcm and consumption about 43 bcm. The remainder was exported to the Russian Federation.

Beginning in 2007, 10 bcm of Uzbek gas will be delivered each year solely under a strategic partnership agreement between the Uzbekneftegaz National Holding Company and Gazprom. Government policy statements indicate that natural gas is widely available throughout Uzbekistan although local media outlets occasionally report supply disruptions in some regions.

While gas supply disruptions could stem from local distribution problems, they could also be due to insufficient gas supplies to meet simultaneously the needs of domestic consumers and the growing exports of gas. If the latter were to be the case, it could prove disturbing to Gazprom having already encountered gas transportation problems moving gas from the Central Asian republics to Russia. For the time being, the Uzbekistan gas transit system is unable to fully handle the transit of 70 – 80 bcm of contracted Turkmen gas to Russia. There is a proposal to build an additional gas pipeline but a number of issues related to the pipeline still need to be resolved including the form of investment in the project.

Russia's Global Energy Security Policy

The Russian Federation has a distinctly different position and role in respect to international energy security than most developed countries, including the other members of the G8 with the exception of Canada. While the other members of the G8, excluding Canada, are major net energy importers, Russia on the other hand is a major net energy exporter with interests that at times may be at odds and competing with the G8 net energy importing countries when it comes to energy matters. On energy, Russia's interests are inherently similar to other major global energy exporters. But, on the other hand, it is a major developed country and member of the G8 and therefore must act as such on global issues, including energy matters. It needs to maintain stable and constructive relations with western net energy importers, such as the United States and the European Union countries, with eastern energy importers, such as China, India, Japan and the Republic of Korea, as well as the other major global exporters of energy.[29]

During the last few years, Russia has achieved significant progress in enhancing its relationship with OPEC as well as the leading member countries of OPEC. In December 2005, the Ministry of Industry and Energy of the Russian Federation established a dialogue between Russia and OPEC and signed an agreement defining the objectives, scope and the general nature of this dialogue. The agreement provides for annual meetings at the ministerial level as well as provides for technical exchanges, seminars and joint research and development.

The establishment of the Interstate Special Energy Group within the framework of the Shanghai Cooperation Organization (SCO) in January 2006 is another important milestone in Russia's quest for international dialogue on energy matters. Since the SCO includes both net energy exporters and net energy importers, the activities of this Group might also be helpful in bridging and balancing the interests of these two groups of countries. In mid-2005, India, Pakistan and the Islamic Republic of Iran joined SCO as observers, further contributing to the reach and influence of SCO.

Russia's stated objective is to be a reliable partner in supplying energy to the world economy. In particular, Russia's aim is to continue to supply the global economy with traditional types of fuels and thereby global energy security. The Government's view is that problems when they arise should be resolved jointly with its international partners, in a way that is acceptable to both producing and consuming states.[30]

Russia intends to expand its hydrocarbon deliveries to global markets, and thereby contribute to the diversification of energy sources by net importing countries. This applies to both countries in Asia and to the United States. At the same time, Russia recognizes that Europe will remain its principal market and a key partner for Russia.

More specifically, Russia intends to take the following measures to contribute to the diversification of markets for Russian and its CIS partner's hydrocarbon exports while, at the same time, contributing to the diversification of hydrocarbon sources for its trading partners (net energy importers):

- large scale penetration of western and eastern Siberian hydrocarbons into the Asian and Pacific markets, including plans for related oil and gas pipelines, terminals and infrastructure. With respect to the Siberian oil pipeline to the Pacific, western Siberian oil is initially expected to cover over half of the pipeline's throughput, with the remainder coming from, and increasingly more over time, from new eastern Siberian deposits, such as Talakan;

- increased natural gas and oil supplies from projects on Sakhalin island, such as that Sakhalin I and II Projects for the Asian and Pacific Region;

- further development of the oil-and-gas transportation network to Europe from Russia and other CIS countries, including expansion of the Blue Stream gas-transit system ; construction of the third line of the oil Baltic Pipeline System, the North-European gas pipeline/North Stream project, and the Burgas-Alexandroupolis oil pipeline; the integration of the Druzhba oil pipeline with the Adria system; and the expansion of the capacity of the Caspian Pipeline Consortium. Construction of the North-European gas pipeline project will open up a major new transport corridor for the delivery of Russian gas to Western Europe. The pipeline, which will be laid along the bottom of the Baltic Sea from Vyborg, Russia to the coast of Germany, will also reduce the number of intermediaries and countries through which the natural gas will have to transit. The pipeline should therefore enhance energy security and the reliability of deliveries by providing an additional transportation corridor and a more direct transportation link between Russia and countries of Western Europe.

- construction of an oil pipeline from the oilfields in the northern European part of Russia to Murmansk, which could allow for the shipment and greater penetration of Russian oil in US and West European markets. According to estimates by Lukoil, the cost of transporting oil from Murmansk to the east coast of the United States could be half as much as the cost of oil deliveries from the Middle East to the Gulf of Mexico. Lukoil and ConocoPhillips have concluded a strategic alliance for developing the northern part of the Timano-Pechora oil and gas Province.

- eventual construction of liquefied natural gas facilities in the Murmansk Region, Yamal Peninsula and/or Barents Sea for shipment to Western Europe and the United States .

Russia is consolidating its dialogue and the coordination of its energy cooperation with other CIS countries that are net exporters of hydrocarbons. The creation of the EurAsian Economic Community (EurAsEC) has contributed significantly to this process.

Russia is seeking to reconcile its national interests with the interests of other members of the global community, both net importers and net exporters of hydrocarbons. But at the same time, Russia would like to reduce its excessive dependence on energy production and exports by developing other sectors of the economy, particularly high-level and mid-level technological industries, that can compete globally.[31] As such, it is seeking cooperative arrangements with other countries that could assist in this diversification of the Russian economy.

As part of this drive to diversify its own energy mix and also to diversify its industrial capacity, Russia has continued to invest in the development of nuclear energy. Estimates by the Government indicate that it would be quite feasible to raise the share of nuclear energy in Russia's electric power generation mix from 16-17 per cent in 2005 to 25 per cent by 2030.

The Government of the Russian Federation has also sought to diminish terrorist threats and provide for greater stability and security in regions beyond Russia, particularly in the area of the former Soviet Union with oil production and transportation infrastructure interconnected with that of Russia, through cooperative arrangements, such as, the Collective Security Treaty Organisation (CSTO), the Shanghai Cooperation Organisation (SCO) and the Commonwealth of Independent States (CIS).

In addition, the Russian Federation has sought to enhance international energy security by promoting economic transition and market-based commercial relations for its hydrocarbon exports. In general, the transition to market-based commercial relations and pricing in contrast to barter schemes leads to greater transparency and predictability in hydrocarbon markets, which ultimately should contribute to enhancing international energy security. Ideally, hydrocarbons should also be sold at market prices in domestic markets. Unfortunately, this may not always be possible for a host of reasons, such as, the inadequate level of disposable income of individuals. However, domestic prices ought to be adjusted and trend towards market-based levels as quickly as possible in line with increases in the purchasing power of consumers. This is clearly an objective of the Russian Government particularly regarding natural gas.

As mentioned earlier, Russian policy is designed to reduce the economy's excessive dependence on energy production and export by developing competitive industries based on high and mid-level technologies.[31] In this regard, Russia is particularly interested in encouraging international cooperation in energy-saving technologies. Although Russia is a major energy exporter, it is also one of the largest energy consumers. This energy intensiveness is not solely due to its geography and its severe climatic conditions. It is also due to the Soviet legacy. The planned economy during the Soviet era created an inefficient energy system especially in industry, agriculture and public utilities. The transition to market-based commercial relations has improved the situation although not sufficiently enough. Russia needs to ensure the widespread deployment of energy-saving technologies and energy management practices that can be best achieved through the formation of markets for energy efficiency.

Therefore, international cooperation notably with the developed net importers of hydrocarbons, may not only help Russia in reducing its domestic energy consumption but may also release additional oil and gas supplies for delivery to international consumers. The energy-saving potential of the Russian Federation is equivalent to about 40 per cent of its current total energy consumption. Therefore, international cooperation in the areas of energy efficiency, renewable sources of energy, and research and development in new technologies for energy production and consumption can have real payoffs for Russia.

1.6 THE NORTH AMERICAN VIEW OF ENERGY SECURITY

Executive Summary

North American energy security faces a wide variety of challenges. Instability in producer countries, natural disasters and infrastructure bottlenecks are just some of the risks to energy supply that have been brought to the fore in recent years. It has also become increasingly clear that one of the best ways to manage the risk to energy security is by developing greater diversification and flexibility throughout the energy supply chain through investments in infrastructure and technology.

The globally integrated nature of energy markets means that investment is needed both within North America and abroad.

Investments in infrastructure and technology can increase available supply by monetizing unconventional sources, creating transportation links to remote reserves and making some sources of energy more environmentally sustainable. Diversification can make supply more reliable and less costly by increasing producer incentives to make supply dependable and affordable. Flexible supply chains can increase overall deliverability by increasing supply options and the ability to respond to unexpected shocks. However, policymakers must provide the frameworks and incentives necessary to mobilize the resources of the private sector and focus the market's efforts on creating greater energy security through investments in infrastructure that provide for diversification of supply.

Contrary to many opinions, the world's energy resources are sufficient to supply global demand for many decades to come. Current proved oil reserves are expected to last at least 40 years and the International Energy Agency (IEA) places the reserve life at closer to 70 years. Despite strong increasing demand for natural gas, proved natural gas reserves are also expected to last at least 60-70 years. Coal reserves in the United States are adequate to meet domestic demand for the next 250 years. In addition to fossil fuel reserves, nuclear power and renewable energy also offer viable sources of increased energy supply. Hence, diverse types of supply are available to meet North America's energy demand.

However, large investments in new technology and infrastructure are necessary to increase the availability of many of these sources. Although the United States has abundant coal reserves, this fuel source will remain controversial unless America's coal fired generation can shift to clean coal technology. The next generation of nuclear power will need strong governmental support to expedite the laborious and costly permitting phase and ensure safety and proper disposal of waste materials. Renewable energy is the most sustainable form of supply but suffers from reliability issues and high costs. Tax credits and demand stimulation, such as requiring that a given percentage of electricity be produced from renewable sources, are necessary to support the investments needed to expand renewable supply capacity.

Supply can also be diversified across sources. North America imports a huge amount of energy, in large part to meet the demand of the United States. Canada, Mexico, Saudi Arabia, Venezuela and Nigeria make up the top five suppliers of oil to the United States. Growing production in Russia, the Caspian Basin, West Africa and Latin America will also provide options for diversifying oil supply. However, as North American refining capacity fails to keep up with demand, the region becomes more dependent on imports of refined products as well as crude. Increasing North America's refining capacity would return control of this element of the value chain to the region and reduce this element of risk. Increased refining capacity, especially in the United States, will only be possible with government support on issues such as more streamlined permitting processes.

With regard to natural gas, as domestic production levels off, the United States and Mexico will both meet a greater proportion of their growing demand for gas with LNG. Atlantic Basin LNG supply options include countries in Latin America, North and West Africa, the Middle East, Northern Europe and Russia. Re-gasification terminals on North America's West Coast would open up the continent to supply from Australia, Southeast Asia and possibly even Alaska. New re-gasification terminals have yet to be built on the strategically important northeastern and western coasts of the United States despite strong efforts; instead new development has clustered in the more receptive Gulf Coast. There is adequate transportation via the Trunkline Pipeline from the United States Gulf Coast to the mid-west (i.e. Chicago, a large demand centre) and to the northeast of the United States. Although there are transportation costs that could be recaptured through the location of LNG receiving terminals in the northeast, where permitting hurdles have been difficult to overcome.

The development of new resources requires enormous investment in infrastructure and technology to enable them to realize their full potential. Energy projects are extremely capital-intensive and require a long time, sometimes a dozen years or more, for the returns on investment to be fully recouped. For example, the cost of the Qatargas II Project, an LNG facility being built in the State of Qatar to supply gas to the United Kingdom, is US$ 9.3 billion. Commitments of this scale are made much more readily if the government receiving the foreign direct investment (FDI) flows can provide a stable foundation (i.e., appropriate and stable legal, regulatory and policy framework) for the investment. Investors must be confident that reliable supply can be produced for the full duration of the project before they will provide funding.

It also needs to be recognized that the relationship between the energy buyer and seller is one of mutual dependency. In most supplier countries, energy sales make up by far the largest proportion of GDP. These countries are also adversely affected by the boom and bust cycle of energy prices and they often attempt to manage this volatility through vehicles such as oil stabilization funds that smooth oil revenues and set aside surpluses for future generations. However, as recent events show, uncertainty and turmoil in producer countries remains. A dispute between Russia and Ukraine disrupted gas supplies to Europe. Despite disagreements between Venezuela and the United States of America, the energy supplies continue to flow, in large part because the sellers need the buyers as much as the buyers need the sellers. Diversification of supply can reinforce that fact by creating competition and putting more pressure on producer countries to be reliable suppliers, or else face the shifting of demand to another region or energy source.

Deliverability is enhanced by investments that link buyers to multiple sources of supply, providing diversification across both supply sources and the transportation infrastructure. LNG provides an excellent example of how investments in infrastructure can increase deliverability. Due to its physical nature, it can be difficult for natural gas to provide diversity in supply or delivery options. Once a pipeline is built, the buyer is locked in to that particular source of supply and all of the risks associated with it. LNG, however, provides flexibility that is not available with pipeline imports. In theory, supply disruptions at one, or even several, supply location(s) would not threaten overall energy security. Disruptions from one supplier can be made up by additional shipments from an alternate source. Similarly, suppliers can have the luxury of multiple potential delivery points and can realize better terms in whichever market is willing and able to offer them. Increasing options throughout the value chain does not lead to insecurity, but rather it can dramatically increase reliability and overall energy security.

Improving the availability, reliability and deliverability of energy supply should lead to better affordability. The current increase in oil prices is not caused so much by scarcity as by threats to the security of energy supply and political tensions in producing countries. Diversification introduces more competition in energy markets, putting downward pressure on prices and increasing the incentive for producers to be reliable suppliers. Greater reliability leads to reduced price volatility and a more stable price environment. Increased investment in infrastructure also creates excess capacity that can be used to provide a supply response to unexpected outages and other supply shocks.

Energy security will not become a reality without governments assuming appropriate roles. Policymakers need to provide a framework that promotes private sector investment in the appropriate infrastructure. For supplier governments this entails providing social stability at home, in part through responsible sharing of energy export revenues. Exporters must also provide stable and fair tax treatment and well-defined property rights that prevent expropriation in any of its various guises. The State of Qatar is an excellent example of a producer country that combines a stable government, sound economic policies and powerful protection of foreign investors' rights to build the strongest LNG export sector in the world.

Consumer governments must provide a streamlined regulatory environment that allows infrastructure such as refineries and re-gasification terminals to be permitted. Subsidies and demand stimulation may be needed to support investment in renewable infrastructure and capture the full value of sustainable energy. Governments must also provide research and development support for new technologies. The United States Department of Energy's FutureGen Program is an excellent example of helping to take a strategic technology, in this case clean coal technology, from the experimental stage to bankable status.

Overall, consumer governments need to promote legal frameworks, which enable the private sector to make investment in infrastructure that diversifies their respective country's energy mix and makes previously unavailable resources available. Investments that create more options for sources of supply and greater transportation flexibility should also be encouraged. For example, in the United States the Federal Energy Regulatory Commission (FERC) has adopted a permitting process for LNG receiving terminals that focuses on safety and transportation issues (including pipeline takeaway capacity), and which does not evaluate or question the market rationale behind the development. This approach allows the market itself to decide which terminals are ultimately the most profitable while also allowing the market to make the investment in additional terminal capacity, increasing delivery options for LNG in North America. Promoting more energy options allows for a stronger response to supply disruptions and enables greater reliability. Increasing diversification increases energy security.

Availability

Oil supply and demand

According to the International Energy Agency's World Energy Outlook, global energy needs are set to increase by more than 50 per cent by 2030, with oil demand rising by about 40 per cent.[32] The main consumers of oil continue to be the advanced economies: the United States, OECD Europe, and Japan, who together consume about half of the world's annual oil output. Consumption in emerging markets, specifically China and India, however, has been increasing at a faster pace. As these economies grow, their energy and oil needs also will increase, further tightening markets and increasing competition for resources.

Oil will remain the dominant energy source even though there may be some downward pressure on demand in the Western world due to environmental concerns and CO_2 emission controls, partial replacement by natural gas, particularly for electricity generation, and a return to favour of nuclear energy in some countries such as the United States and the United Kingdom. In the United States, as shown in Figure 1.6.1, the continued dominance of oil is overwhelmingly driven by the transportation sector, which accounts for 75 per cent of oil consumption in the United States.

Figure 1.6.1 United States Oil Consumption by Sector, 2004

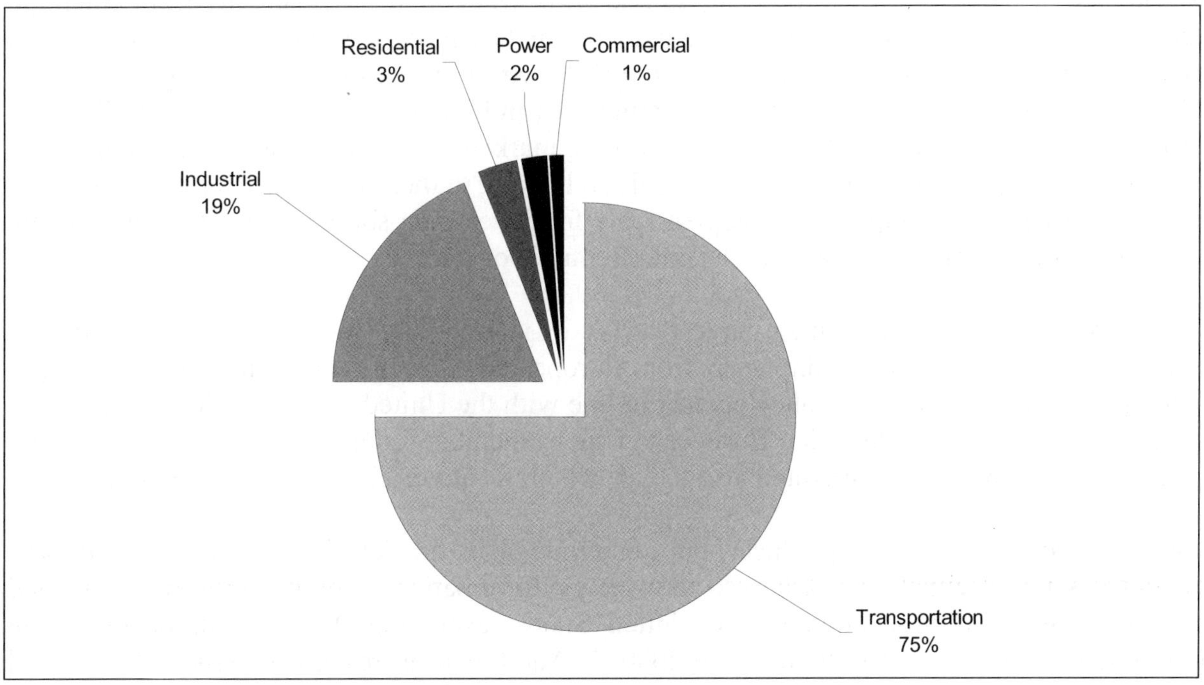

Residential 3%
Power 2%
Commercial 1%
Industrial 19%
Transportation 75%

Source: United States Energy Information Agency (EIA)

Proven oil reserves, according to BP, are sufficient to meet global production forecasts at current levels for over 40 years.[33] This figure, however, significantly underestimates the volume of oil resources that may eventually be recoverable with improved technology or at higher oil prices. On this basis, the IEA calculates that remaining oil resources could be extended to 70 years at average annual production projections from 2004 to 2030.

Although non-Middle East suppliers will provide a more substantial volume of oil in the near to mid-term, the primary suppliers will remain OPEC member states, whose share will approach 50 per cent by 2030.[32] Gulf member states have developed their respective oil sectors largely independent of international oil companies (IOCs). The IOCs have been investing principally in fields in the North Sea, Alaska, West Africa and the Caspian basin.

Non-OPEC developing countries are expected to increase their production volumes over the period to 2030, partly as a result of enhanced exploration and extraction technologies. In particular, oil producers in Central and South America have significant potential to increase output over the next two decades. So do producers in West Africa. Production in Russia and from countries in the Caspian region is also expected to rise. On the other hand, production from OECD countries is projected to decline over time. Over the period to 2030, total non-OPEC production is expected to increase by about 0.7 per cent per year in contrast to 2.1 per cent per year for OPEC production.[32]

The current situation will change gradually if Middle Eastern OPEC member countries begin the process of opening up to more foreign direct investment and new technology. At the same time Russia and the Caspian basin will continue to contribute to available oil resources, as will Latin America and West Africa; oil production in these regions is expected to increase by an average of 1.5 per cent a year until 2030.[96] In addition, non-conventional production technologies, such as for the oil sands in the Canadian Province of Alberta and the extra-heavy crude oil in the Orinoco Belt in Venezuela, will provide production gains, particularly given that oil prices are making these methods economically viable. Most estimates put the minimum oil price at US$ 40 per barrel for unconventional oil sources to become a viable source of supply.

The worldwide demand for oil will continue to rise with strong demand emanating from emerging markets, adding to the already increasing demand from mature markets such as those of the United States and the European Union. According to the IEA, world oil consumption is projected to rise by 1.3 per cent per year, with consumption reaching 99.3 million b/d in 2015 and116.3 million b/d in 2030 compared to 83.6 million b/d.[32] The emerging markets, including notably India and China, are leading the rise in demand. These markets will be looking to their traditional supply sources for oil and gas, but will also compete more aggressively for developing sources around the world and, in particular, in the Caspian Basin and the Islamic Republic of Iran.

The United States, which is the single largest driver of international markets, will be faced with increased competition over the coming years from Europe and Asia. It remains to be seen if Europe will continue to follow an energy policy agenda in line with the United States or whether there will be sufficient consensus within the European Union member countries to develop a more independent approach to oil security and also to gas, which would result in greater competition.

The United States is a major oil producer, but numerous of its oil fields are nearing maturity and production is set to decline. According to the Energy Information Administration of the United States Government, oil production in the United States, excluding Alaska, will decline from 2.9 million b/d in 2004 to 2.3 million b/d in 2030.[34] Additional oil resources exist in Alaska and offshore in the Gulf of Mexico which are not currently exploited. Unconventional oil sources such as oil shale have also not yet been tapped since they are not economically viable using current technologies.

The United States, despite being a major oil producer, has large energy needs, which will continue to be met by its main providers; Canada, Mexico, Saudi Arabia and Venezuela (see Table 1.6.1). In 2005, the United States produced 1.9 billion barrels of crude oil and imported 3.7 billion barrels.[35] Canada's recent development of its oil sands bodes well for the United States. These resources, however, will not stem the risk associated with dependency on traditionally politically risky countries. China's ventures into Venezuela and Nigeria's current social unrest pose a serious threat to United States oil supply.

Table 1.6.1 Top Ten United States Crude Oil Suppliers, 2004 (b/d)

Country	**Thousand Barrels per Day**
CANADA	1 868
SAUDI ARABIA	1 457
MEXICO	1 576
VENEZUELA	1 169
NIGERIA	1 075
ANGOLA	379
ALGERIA	350
IRAQ	666
RUSSIA	255
ECUADOR	239

Source: United States Government, Energy Information Administration (EIA)

United States imports of oil are expected to increase due to the fact that domestic production is unable to keep up with the predicted growth in crude oil consumption.

Figure 1.6.2 United States Crude Oil Production and Consumption, 2003-2030

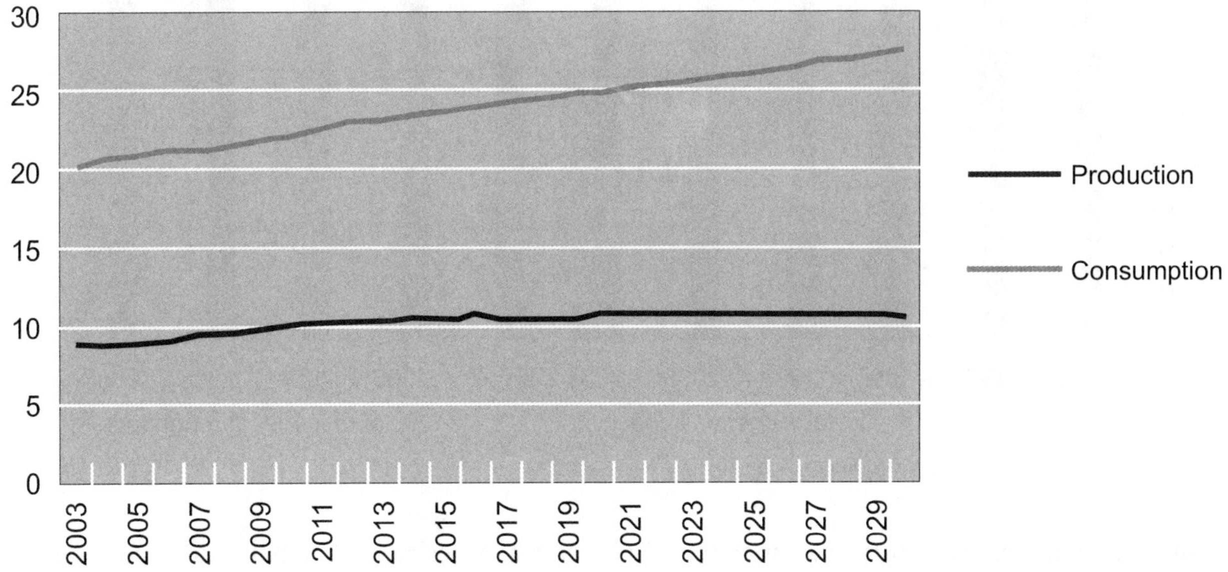

Source: United States Government, Energy Information Administration (EIA)

Oil Refining Capacity

The rapid global growth in demand for refined products over the past several years is contributing to a reduction in spare capacity. As seen in Figure 1.6.3, capacity growth is not keeping up with demand for greater throughputs. Without additional capacity, it will not be possible to meet the burgeoning demand for products, namely gasoline and diesel. Despite economic and regulatory hurdles, a modest increase in investment in new distillation and upgrading capacity, particularly in the Middle East and developing Asia, is foreseen over the next couple of years. However, it would appear that the international product market will remain tight over the next decade and possibly beyond. The IEA estimates that utilization rates will increase to more than 86 per cent by 2010 compared to the more normal 80 per cent.

One of the major factors contributing to diminishing product capacity is increasingly stringent sulphur regulations. The lower sulphur targets for gasoline and distillates are resulting in great competition for lighter and sweeter crude oils, increasing the price spread between the sweet and sour crude oils. Europe has led the way with a maximum sulphur content restriction of 10 parts per million by 2009.[32] India and China, among many other countries of the world, also are tightening their specifications. Refiners have been presented with one of two options: invest in new units - coking, cracking or hydro treating facilities to upgrade their products - or purchase the more expensive crude.

Figure 1.6.3 Total Refinery Capacity and Throughputs, 1996-2006 (thousand b/d)

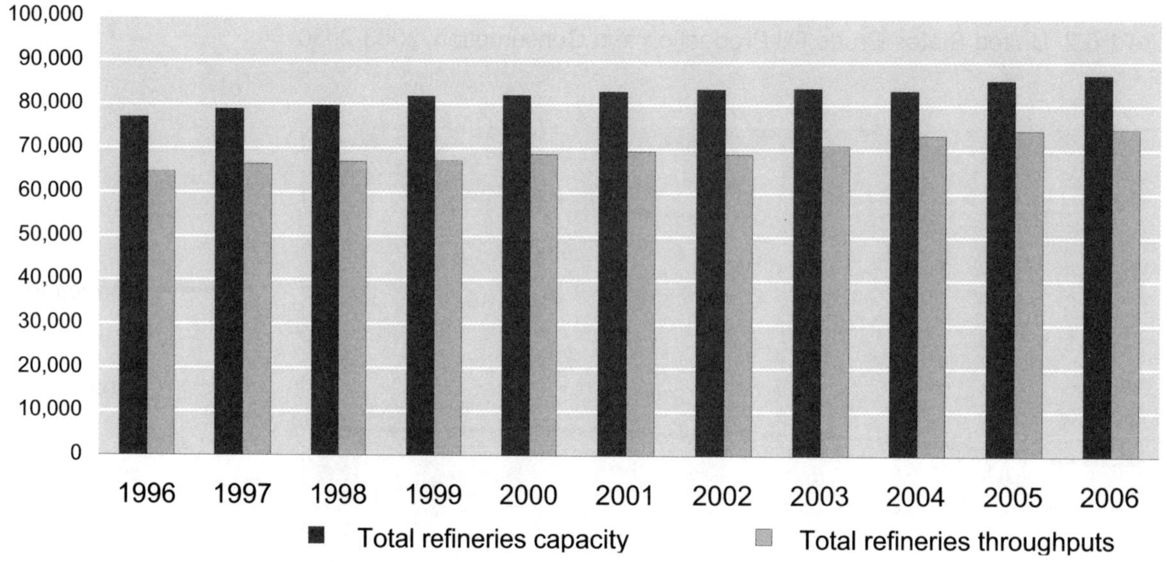

Source: BP Statistical Review of World Energy June 2007

Despite minor increases in refinery capacity, OECD North America will remain a net importer of refined products. In 2005, United States refinery capacity increased to 17.1 million b/d from 16.5 million b/d in 2000, an increase of 0.6 million b/d.[36] United States demand for refined products, however, grew at a more rapid rate than capacity, increasing to 21.0 million b/d in 2005 from 19.7 million b/d in 2000, an increase of 1.3 million b/d.

The major suppliers of the United States market as well as those of Canada, the United States Virgin Islands and Europe, will be able to meet United States gasoline specifications. Refiners in South America and other regions, however, face greater difficulty as they lack the necessary units to upgrade their products and are faced with greater domestic gasoline demand growth than the United States. This can prove problematic as demand for gasoline and distillates in the Far East and Latin America can potentially pull products from the United States and from its key suppliers; Asia Pacific and Latin America refined product consumption is growing at a faster pace than in the United States.[33] Moreover, Europe, coping with new environmental regulations, is projected to become a net importer of middle distillates by 2010. To ensure adequate supply, investment in refinery capacity additions and downstream distribution systems is required.

The IEA estimates that US$ 770 billion of refinery investment (in 2005 US dollars) will be required to keep up with global demand grow and for meeting tighter environmental standards for the period 2005-2030.[32] With regard to total energy infrastructure investment, an overwhelming percentage, about two-thirds, of the refinery investment will be needed in emerging markets with the majority of this investment earmarked for servicing their domestic refined products market.

The United States Government attempted to promote investment for expanding and developing refineries in the 2005 Energy Bill. This Bill allows for the expensing of 50 per cent of the investment costs that increase capacity by at least 5 per cent or that increase throughput of oil from shale and tar sands by at least 25 per cent. So far, there are no signs that this provision has or will lead to much higher investment in refinery capacity, partly because capital and operational expenditures continue to be lower abroad, particularly in Asia and the Middle East.

Building a large greenfield refinery in the United States is both capital-intensive (i.e. the starting price is in the order of US$ 5 billion) and time-intensive over the development, permitting and construction phases. Although current margins are strong due to deflated surplus capacity, the refining sector has seen weak returns for the past 30 years; whereas integrated oil companies would expect returns in the order of 20 per cent for investments in the upstream sector, refiners and marketers have been earning less than 10 per cent.

Unless substantial expansive investments are made, the United States will be more exposed to supply disruptions as its share of refined product imports increases. This risk is further exacerbated by the construction of refineries close to crude oil supply in areas such as the Middle East, Africa and Venezuela. In the event of a disruption or crisis in one or more of these locations, the United States will face not only a shock to crude oil supply, but a shock to refined product supply as well. Even if replacement crude oil can be found from alternative sources, without corresponding refining capacity the replacement supply is of marginal benefit. In order to manage this risk, it would be helpful to streamline the permitting process in order to reduce the development phase of refinery investments. This would lower the cost of potential projects and increase the chance of additional refineries being constructed in a timely manner.

Natural gas supply and demand

Natural gas has seen a dramatic period of expansion, a trend that is likely to continue. According to the IEA, natural gas will continue to be the fastest growing source of energy with its market share increasing from 21 per cent in 2004 to 23 per cent by 2030, mostly at the expense of nuclear energy but also to some extent coal.[32] This growth has been and will continue to be overwhelmingly driven by power generation demand.

Three countries, the Russian Federation, the Islamic Republic of Iran and the State of Qatar, account for two-thirds of the world's gas reserves. The State of Qatar has been more aggressive in attracting commercial development of its fields than the Islamic Republic of Iran by encouraging FDI in large scale LNG, petrochemical and gas-to-liquid (GTL) developments. As a consequence, annual flows of FDI have risen from US$ 34 million in 1995 to US$ 679 million in 2004.[37] Total GDP rose from US$ 17.8 billion in 2000 to US$ 28.4 billion in 2004 and Qatar's GDP per capita has grown from US$ 29,290 in 2000 to US$ 36,620 in 2004.[38]

Given the anticipated growth in energy and particularly gas demand, there have been considerable efforts to explore for gas and increase capacity from already known fields. Development to date has been fairly uneven with much of the progress being concentrated in the North Sea, the United States and the Russian Federation, while the Middle East has lagged behind. However, this situation is changing as Qatar develops some of the word's largest gas export projects, notably Qatargas II and RasGas III and IV. Over the longer term, countries of the Middle East, Africa and developing Asia will supply increasing amounts of natural gas as new fields and export projects come on line.

According to the IEA, natural gas production from the Middle East and Africa is expected to grow at an average annual rate of about 4.5 per cent respectively for the period 2004-2030 compared to 0.4 per cent for North America and –0.5 per cent for OECD Europe.[32] The countries that will play the most significant role as exporters will be Qatar, Iran, Algeria and Saudi Arabia (some of the gas exports will be via petrochemicals and plastics). The Middle East will shift its trade towards the West with new gas production meeting the gap in Europe caused by depleting regional supplies in the North Sea. Europe will become the primary market for North African gas, particularly with the construction of new export pipelines to Italy and Spain.

The growth in LNG is booming. Rising energy costs and decreasing costs of production combine with a whole new market paradigm where projects are no longer being developed as "point-to-point" solutions, but rather as supply available to multiple markets.

All net importing countries are facing greater energy consumption needs. The European Union will be importing 80 per cent of its gas by 2030, according to the IEA. North America, which was traditionally self-sufficient in terms of natural gas, will import 14 per cent of its gas needs in 2030.[32]

North American production capacity in recent years has levelled off with existing import sources no longer able to meet increased demand. Demand is set to rise 1.2 per cent per year, with two thirds of the growth going to the generation of electricity, over the next two decades. The market will remain tight until new supply from Alaskan sources, Canadian fields and increased LNG imports can be brought online. The tightness of the market explains the surge in prices that can be seen in Figure 1.6.4. Political tensions following the tragedy of 11 September 2001 in New York and Washington as well as weather related incidents along the Gulf of Mexico have also contributed to the increase in prices.

Figure 1.6.4 Price of Natural Gas in the United States, 1990-2005

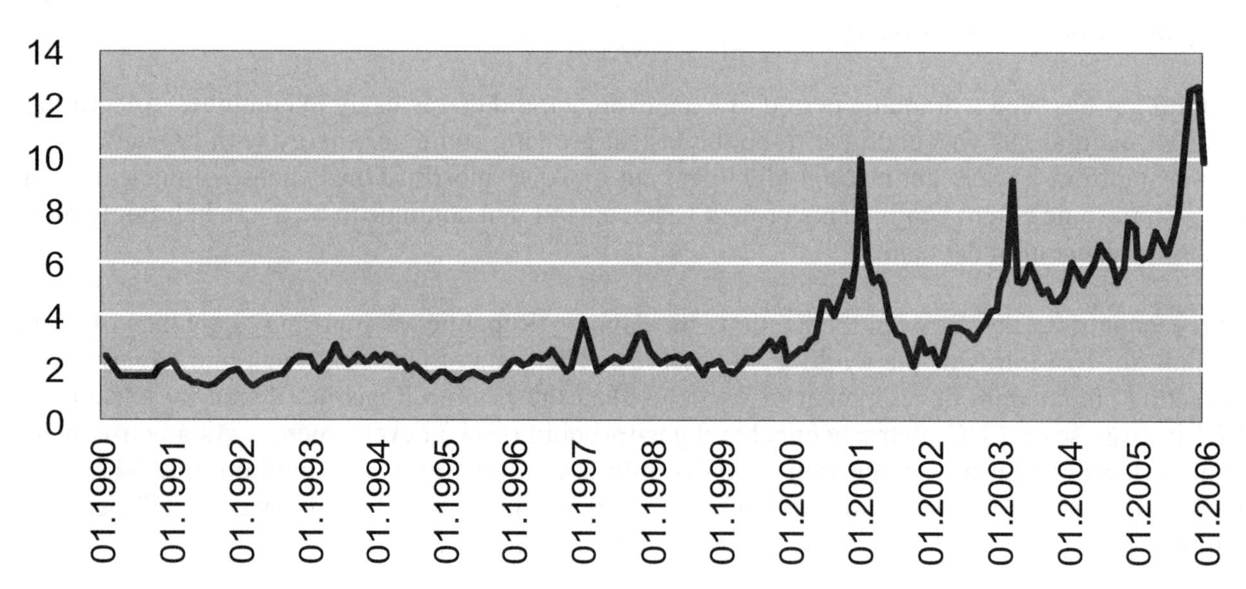

Source: Chemical Market Associates, Inc. (CMAI)

In Mexico, natural gas consumption is expected to outstrip production. Mexico's demand for natural gas is projected to grow at an average annual rate of 3 per cent between 2002 and 2025, while production is expected to grow at a rate of only 1.7 per cent annually. As in the United States, most of the growth in consumption is expected to fuel electricity generation.

Despite the difficulties gas presents as a source of energy due to its high initial costs and complexity to transport and store, a burgeoning demand for energy and increasing economies of scale make natural gas a competitive commodity.

Developing New Technologies to Improve the Recovery of Fossil Fuels

According to the IEA, the world contains approximately 20 trillion barrels of oil equivalent (boe) of oil and gas, of which only 5 to 10 trillion are recoverable today.[39] One of the greatest challenges to improve energy security is thus to enhance the recovery of oil and gas reserves.

New tertiary recovery or enhanced oil recovery (EOR) technologies are being developed to address this challenge. These techniques include thermal recovery through the injection of steam, chemical injection of polymers and gas injection, usually CO_2. There is much interest in EOR techniques. The United States Department of Energy has an extensive EOR research programme. According to a 2004 Department of Energy report, CO_2 injection could enable the United States to recover an additional 43 billion barrels of currently stranded oil.

The United States Department of Energy is also active in promoting research in other technologies that could improve oil production in the United States. These technologies include better diagnostic tools to more accurately evaluate reserves. Since 78 per cent of United States oil fields are nearing the end of their lives, the Department of Energy is also paying special attention to marginal well revitalization. Marginal wells are oil wells whose production has slowed to less than 10 barrels a day and whose exploitation is, as a consequence, no longer commercially viable. However, significant quantities of oil do remain in these wells, which could be recovered using proper technologies.

Oil sands and oil shale projects were developed in the United States and Canada following the 1973 global oil price hikes but interest subsequently waned during the 1980s when oil prices sharply declined and the economic viability of these energy sources became questionable. Today's revival of interest in the oil sands in Canada is linked to similar conditions as those in the 1970s; high crude oil prices, tight supply, and the fear that the Western world is running out of oil. Abundant non-conventional oil reserves in politically stable countries such as Canada seem attractive, amplified by improved technologies that make their recovery commercially viable in a world where crude oil prices are likely to stay above US$ 30 per barrel. Alberta's oil sands production in Canada is booming and is expected to reach 5 million b/d in 2025.[40] Approximately US$ 90 billion of investments are planned in the oil sands over the next two decades and the industry expects positive returns for oil prices in the range of US$ 30-40 per barrel and perhaps as low as US$ 21 per barrel.[40]

Canada's bitumen reserves are not the only source of synthetic crude. Venezuela has led the way in the exploitation of ultra heavy crude oil and is planning large developments in new fields in the Orinoco Belt. Reserves in this area amount to approximately 1.3 trillion barrels, but according to Petróleos de Venezuela S.A. (PDVSA) only 120 million barrels are currently recoverable.

Other unconventional oil and gas sources remain but they are far from the commercial stage of development. The United States has the world's largest reserves of oil shale; oil shale refers to a rock containing bituminous substances that can be extracted by heating the rock. United States oil reserves could be enhanced by the recovery of between 1.5 and 2 trillion barrels from these oil shale deposits.[41] Oil shale projects were developed in the 1970s, but abandoned in the 1980s because they were not commercially viable in the context of lower oil prices. The rise in oil prices since 2003 has spurred a renewal of interest, but current technologies do not allow extraction at cost competitive prices; oil shale could become viable only with a price per barrel of at least US$ 70 to US$ 95.[41] Shell is conducting experiments on a new technology that could have promising results, but not before the next decade.

Unconventional gas sources are also being studied; most of them associated with or derived from coal. Another unconventional gas source is methane hydrate; ice crystals in which methane molecules are trapped. Methane hydrate can be found on land, in permafrost regions and under the sea at water depths below 500 metres. World methane hydrate resources are very large, literally dwarfing current natural gas reserves. No technology currently exists, however, which would enable oil and gas companies to exploit these resources.

Coal

As a result of increased concern for energy security, improved clean coal technologies and the presence of abundant reserves, including in the United States, coal has received new attention recently as an important energy source for power generation. Of all the fossil fuels, coal has the largest reserves worldwide. The United States alone has reserves that will last for another 250 years. The United States consumed 1.1 billion short tons in 2004 of which 20 million short tons (less than 2 per cent) were imported; nearly all of the coal imported by the United States in 2004 came from Colombia. At current production levels, global proven coal reserves are estimated to last 147 years, compared with 41 and 63 years for proved oil and gas reserves, respectively.[33]

Coal plays a significant role in power generation, providing fuel for 40 per cent of the electricity produced globally. Coal fuels about 50 per cent of the power plants in the United States.[34] Coal use in power generation is projected to grow by 60 per cent in the period to 2030. Increasing prices for natural gas combined with projections of relatively stable coal prices are the key factors helping to rejuvenate the coal industry. Currently, over 6,190 Mt of coal is produced globally – a 60 per cent growth over the past 25 years. Coal production has grown fastest in Asia, while Europe has seen a decline in production. The largest coal producing countries are China, the United States, India, Australia and the Russian Federation (see Figure 1.6.5). Much of the global coal production is used for domestic consumption; only about 18 per cent of hard coal production is destined for the international coal market.

Global demand for coal is projected to rise by 1.8 per cent per year over the period to 2030.[32] Power generation accounts for most (about 80 per cent) of the projected growth in coal consumption globally. In fact, the power sector's share of global coal demand will rise from 68 per cent in 2002 to 73 per cent in 2030. Strong growth in electricity consumption is expected in developing countries, where electricity demand is forecast to increase by an average of around 4.0 per cent per year.

Figure 1.6.5 Top Ten Coal Producing Countries, 2004 (Mtoe)

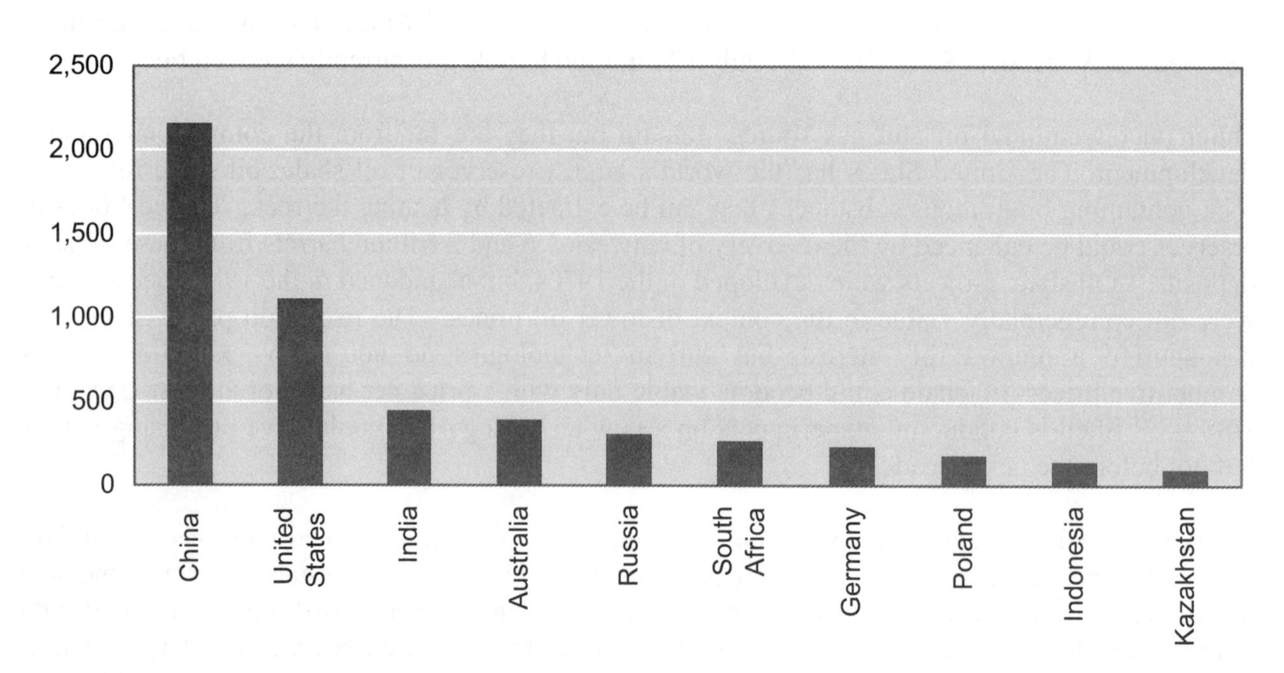

Source: United States Government, Energy Information Administration (EIA)

The Asian economies are continuing to grow rapidly, with India and China leading this growth. Coal currently provides around 78 per cent of the electricity supply for China and 69% for India.

In the United States, coal-burning power plants produce about half of all electricity generation. Rising natural gas prices, which has made some gas-fired plants too expensive to operate, is also contributing to the increase in coal demand. By 2025, coal's share of total electricity generation in the United States is expected to grow from 50 to 53 per cent.

Figure 1.6.6 Top Ten Coal Consuming Countries, 2004 (Mtoe)

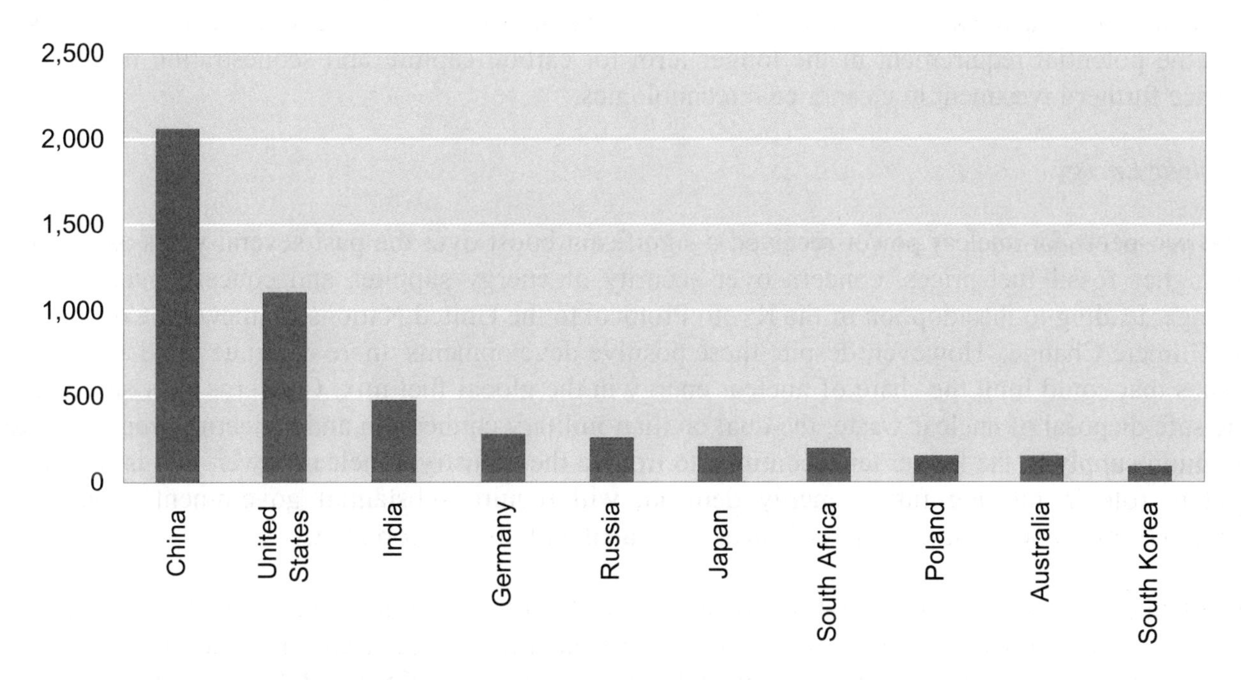

Source: United States Government, Energy Information Administration (EIA)

As a result of rapid expansion, the United States will need to boost its existing coal-fired power plants by nearly 90 gigawatts (GW) of greenfield capacity by 2025.[34]

In order to realize coal's potential, however, coal power systems will have to meet increasingly stringent regulations. Emissions from coal-fired power plants include sulphur dioxide, mercury, nitrogen oxides and carbon dioxide. Combusting coal in traditional power generating and heating plants releases about 8 billion tonnes of CO_2 globally into the atmosphere each year, including 2 billion in the United States.

With technological advances, new coal power systems are already dramatically cleaner and more efficient than plants based on older technologies. Current approaches include further development of combustion systems and deployment of technologies utilizing combined cycles, such as supercritical pulverized coal combustion (SCPC), circulating fluidized bed combustion (CFBC) and integrated gasification combined cycle (IGCC). For zero or near-zero emissions, carbon dioxide will need to be captured and permanently stored. IGCC technology is currently the only coal-based power system capable of capturing carbon dioxide, the efficiency rate for the production of electricity is about 50-60 per cent compared to about 33 per cent for a traditional coal-fired plant and the CO_2 content of the off-take gases is sufficiently high for CO_2 capture.

In 2003, the United States Department of Energy announced the FutureGen initiative. This is a cost-shared partnership between the Federal Government and an alliance of coal-utility and coal-producing companies that will construct the world's first zero-emission coal gasification power plant. Its total cost is estimated at about US$ 950 million, with US$ 250 million of that coming from the private sector. The FutureGen plant, operating at a nominal capacity of 275 MW, will produce both electricity and hydrogen and will sequester one million metric tons of CO_2 annually. A site selection process was initiated in 2006 and operations are scheduled to commence in 2012. It is hoped that the technologies developed and proven during this project will result in the construction of similar plants in the United States and internationally.

The Energy Policy Act of 2005 in the United States provides significant federal financial incentives for a variety of gasification projects. The expectations of more stringent environmental regulations and the potential requirement in the longer term for carbon capture and sequestration might also induce further investment in cleaner coal technologies.

Nuclear Energy

The prospects for nuclear power received a significant boost over the past several years as a result of higher fossil fuel prices, concern over security of energy supplies and concern over climate change, leading to the adoption of the Kyoto Protocol to the United Nations Framework Convention on Climate Change. However, despite these positive developments, there continue to be a range of issues that could limit the share of nuclear energy in the global fuel mix. Concerns over safety and the safe disposal of nuclear waste, the dual civilian-military connection and concerns over adequate uranium supply in the longer term continue to trouble the industry. Nuclear power, if it is to play a greater role in meeting future energy demand, will require substantial government support in underwriting some of the commercial risks associated with its development.

The United States has more operating nuclear power facilities than any other country in the world, with 103 commercial reactors. These supply 20 per cent of the electricity produced in the United States. Although no new reactors have been built since 1996, the Energy Policy Act of 2005 seeks to enable future development of nuclear power.
Nuclear energy plants are dependent on the availability of uranium as a fuel source,. The current uranium requirements of the world's nuclear power reactors is approximately 67,000 tonnes per annum. According to the Red Book, prepared jointly by the OECD's Nuclear Energy Agency and the United Nations International Atomic Energy Agency, the world's present known economic resources of uranium, exploitable at below US$ 80 per kilogram of uranium, are about 3.5 Mt, which, at today's usage rate, is enough to last for 50 years.

Table 1.6.2 highlights the current availability of uranium resources. Australia possesses the overwhelming majority of supply, with Kazakhstan a distant second.

Table 1.6.2 Known Recoverable Resources of Uranium

Country	Thousand Tonnes	Per cent of Global Supply
Australia	1 074	30
Kazakhstan	622	17
Canada	439	12
South Africa	298	8
Namibia	213	6
Russian Federation	158	4
Brazil	143	4
United States	102	3
Uzbekistan	93	3

Source: World Nuclear Association

While risks to uranium supply are not of the same nature and magnitude as those for hydrocarbons, there are nonetheless some risks. Political obstacles related to uranium as a source of nuclear power, long lead times for the opening of new mines and transportation difficulties are among the problems that could limit growth of the nuclear energy industry.

Figure 1.6.7 Top Ten Net Nuclear Electric Power Consuming Countries, 2004 (billion kWh)

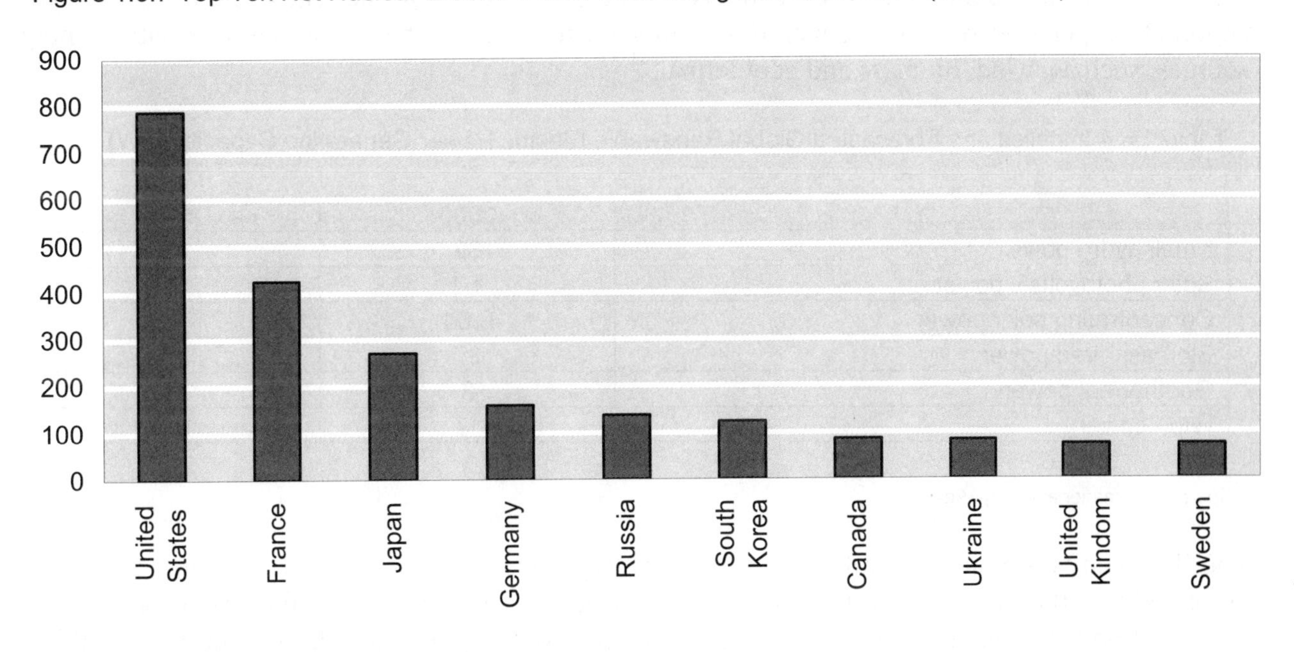

Source: United States Government, Energy Information Administration (EIA)

Investments in nuclear power require stronger government support than investments in most other forms of energy. Nuclear power plants not only require support to underpin their economic viability, but also need government assistance in ensuring the safety of plants, disposing of radioactive waste and securing the technology to prevent proliferation.

Renewable Energy

A 2004 IEA study reviewed data from 1970 to 2001 and found that over this period renewable energy's portion of primary energy supply of IEA member countries had increased slightly from 4.6 per cent in 1970 to 5.5 per cent in 2001.[42] The table below shows the relative growth of each renewable category over the study period.

Table 1.6.3 Average Annual Growth Rates of Renewable Energy Sources (per cent)

	1970-1980	1980-1990	1990-2001
Renewables	3.2	2.4	1.2
Biomass	3.5	3.0	1.6
Hydro	2.6	0.7	0.4
Geothermal	8.3	9.4	0.4
Wind/Solar	6.4	23.5	23.1

Source: International Energy Agency

Table 1.6.3 shows that the strongest growth took place in the wind and solar category. Although these high growth rates are in part due to the fact that the growth is from a low base, it also reflects the strong policy support that these categories have received over this time period. More recent reports indicate that these strong growth rates have been sustained or exceeded over the past several years as well, especially in the wind category.

The role of renewable energy is strongest in the supply of electricity. A separate IEA report published in 2003 found that in 2000, renewable energy provided 19 per cent of global power generation in 2000, second only to coal (39 per cent). Nuclear, natural gas and oil provided 17 per cent, 17 per cent and 8 per cent, respectively, of the global total. In the renewable category, hydro supplied 92 per cent of power generation compared to 8 per cent for the new alternative energy sources, such as, wind, biomass and geothermal.[43]

Table 1.6.4 Installed and Forecasted Global Renewable Electric Power Generating Capacity (GW)

	2000	2010
Small hydro power	32	45
Solar photovoltaic power	1.1	11
Concentrating solar power	0.04	2
Biomass-fired power	37	55
Geothermal power	8	14
Wind power	17	130

Source: International Energy Agency.

The IEA study examined several types of renewable power, shown in Table 1.6.4. Strong growth is projected for all of these renewable classes, with particularly strong growth in the solar photovoltaic and wind power categories. It is important, however, to recognize that despite the projected high growth rates, biomass, wind, solar and geothermal sources are not likely to contribute more than 5 per cent to global electricity generation over the coming decade and probably no more than 10 per cent by 2030 assuming no additional government support. Even the projected growth would not be possible without continued support from government policies promoting investment in renewable sources of power.

Currently, most investments in renewable power generation need some form of government support to be viable. Although under optimal circumstances some forms of renewable generation can compete with conventional sources of power, optimal conditions do not frequently exist. Often the sites suitable for renewable development are far from the grid and thus entail large investments in transmission along with generation. Technologies are often new and investors may require robust project economics to underwrite technology risks. Most of these industries are small and need support that will enable manufacturers to reach a size where economies of scale enable renewable generation to be more competitive.

Government support has been provided in many different forms. Direct payments to producers and more indirect subsidies such as tax credits have been enacted. Policies that direct utilities to provide a certain amount of power from renewable sources, as well as set the prices at which renewable power must be purchased, have also increased the incentive to invest. Finally, taxes on conventional power have also attempted to shift consumer demand to greener sources of electricity. These types of policies have been applied in different countries and categories of renewable power with a variety of effects.

Wind-powered electricity generation has grown rapidly over the past several years, and this growth has been due in part to strong public policy support. In the United States, the Production Tax Credit (PTC) has largely fuelled the growth in wind power generation. The PTC provided a 1.5 cents per kWh production tax credit for ten years to wind power generation built between 1994 and 1999. It was renewed in 1999, 2001, 2004 and 2005.[43] The PTC was instrumental in making wind power economical and in increasing the amount of wind generation capacity in the United States. However, the uncertainty surrounding the renewal of the tax credit created a boom and bust cycle in the United States wind industry. Following a PTC renewal, a large amount of investment in wind power would take place, with investment drying up as the credit neared expiration. This environment has inhibited the development of the domestic wind industry, especially in manufacturing where producers are unable to forecast demand with any certainty. The stability of the German and Danish policies has provided greater opportunity for growth.

One policy that has proved effective in the United States at promoting solar photovoltaic generation is net metering. Net metering allows producers of renewable energy to sell surpluses back to the grid. This provides smaller producers (residential or small commercial entities) with an incentive to invest in their own generation capacity. When conditions are optimal and they are producing more power than they need, they can sell that power to the grid and use the earnings to offset the cost of their investment.

Overall, there are a number of ways in which governments can provide support to renewable power generation, including direct subsidies, tax credits, and fixed off take prices. All of these policies are important in making renewable projects cost-competitive and economically viable. However, just as important is policy stability. Investors need to know that the support that is in place today will be available tomorrow in order to formulate long-term investment strategies and create strong, sustainable growth within the industry.

Reliability

There are a number of significant risks that can threaten energy supply and energy security. These risks include political risks, the concentration of supply among a small number of producers, force majeure and infrastructure bottlenecks. As supply becomes concentrated in a smaller number of producer countries that also exhibit high levels of political risk, the overall risk to energy security also increases. It therefore becomes increasingly important to build out the infrastructure necessary to maximize diversity of supply to enable better management of the political risks. Increased investment in energy infrastructure also builds flexibility into the system, allowing better responses to force majeure events and decreasing the occurrences of infrastructure bottlenecks.

Energy investments are extremely capital-intensive and long-term in nature. Costs can run into the billions of dollars and require a dozen years or more for returns on investment to be recouped. Commitments of this size are not made lightly, and they become more difficult to make in an environment characterized by turmoil and instability.

One fact that needs to be recognized, however, is that the relationship between energy buyer and seller is one of mutual dependency. The economies of most producer countries are heavily dependent on revenues from energy exports, and therefore they are as equally interested in providing reliable supply as the buyers are in receiving it. This fact provides some leverage to importing countries. By diversifying supply options, buyers can create more competition between sources of energy, thus placing more pressure on exporting countries to be reliable partners.

Political Risks

Political risks in traditional supplier countries are hindering the ability to finance needed production capacity projects. While export credit agencies and international financial institutions such as the World Bank increasingly are providing loan guarantees that address issues of force majeure and political instability, there are no guarantees for actual product supply. Exploration and production (E&P) ventures in volatile climates have been encouraged through the advancement of financial instruments that protect investors, but political risks to supply remain.

Allocation of oil revenues is a politically sensitive topic. While it is important to reinvest in exploration and production to ensure adequate reserves, it is also important to pass on some of the benefits to citizens through social or other programmes. The inability to balance these competing interests can lead to the undermining of the domestic oil and gas industry. Civil unrest in countries such as Angola, Nigeria and Iraq has led to the sabotage of production facilities and pipelines. Inhabitants of oil-rich countries who are dissatisfied with their leaders – be it due to disputes over land rights, resource royalty payments or the marginalization of ethnic or socio-economic groups – have a clear target through which to strike at the government establishment. Energy security in the current system will depend on greater cooperation and understanding between governments, and greater fiscal and political reforms in oil producing countries. Issues of transparency and government corruption must be addressed to settle civil unrest and to attract and maintain private investment, and fiscal reform will be needed to improve reinvestment in national petroleum companies.

Turning to Latin America, due to its geographic proximity this region is and has been a traditional energy supplier for the United States. Mexico is the second largest source of United States oil imports after Canada, and Venezuela is the fourth. Latin American energy resources were seen by the United States, as well as the European Union, as a way to diversify away from Middle East dependency. Significant investments were undertaken in the upstream sector, but the resurgence of political tensions in the area could stem this flow of investment.

Political risk involving oil and gas supply from Latin America largely centres on political disagreement and discord with the United States that has characterized some of the governments that have come into power over the last few years. There is increasing political tension due to strengthened ties between Venezuela and other countries that have similar differences with the United States. Likewise, the growing importance of Chinese interest in Latin America is also adding to concerns about increased competition and the potential diversion of oil supplies to Asian countries with fast growing economies.

An additional security concern is that national oil companies such as those in Mexico and Venezuela have been plagued by under investment, which has implications for production and supply. The ability of Mexico to maintain its projected capacity has been further undermined by recent re-evaluations of reserves that place them at levels well below previous estimates. In terms of ability to supply the United States market, Mexico is faced by the same production problems as the United States in the Gulf of Mexico. The United States is faced with a supplier that can be closed down by hurricanes at the same time as its own domestic production is jeopardized.

In the Republic of Bolivia, there is the tension of renewed interference in national oil production and calls for nationalization. Evo Morales, who was elected President in December 2005, has pledged to nationalize the petroleum sector and redistribute privately owned land to the poor. Should President Morales fully follow-through with his promises, there could be consequences for international private petroleum companies in terms of loss of assets and investment. With the reputation of national petroleum companies for under-investing revenues in exploration and production, nationalization could also hurt gas supplies to the international market.

Sub-Saharan Africa continues to be an important supplier of oil and coal and, more recently, LNG to both Europe and the United States. However, the major African producing regions are plagued by political and economic instability, as well as corruption causing concern over the region's reliability as an energy supplier.

Nigeria recently has been a source of unrest where violence and protests have targeted foreign oil companies, their employees and infrastructure, undermining continued investments to monetize energy resources. The targeting of oil companies is due to the fact that oil revenues are perceived not to sufficiently benefit local inhabitants, coupled by the failure to provide basic public needs and employment. Corruption is also a concern. These issues have given rise to a rash of attacks that have plagued the country and at times cutting Nigeria's exports of 2.5 million b/d by nearly 10 per cent. Many of the problems seen in Nigeria are not unique to the country.

Angola is another oil-rich country in the region and it may also soon become an LNG supplier through its Angola LNG project. The country is, however politically uncertain in terms of its general political situation. After achieving an uneasy peace in 2002 following numerous years of civil war, the country is now awaiting legislative elections to take place in 2008 and a presidential vote in 2009.

The Republic of Equatorial Guinea, another important African oil and gas supplier, is faced with political instability after several alleged coup attempts in its recent history. The country, however, is about to become an important source of LNG; the 3.4 million tonnes per annum Equatorial Guinea LNG Project, scheduled to come online in 2007, is contracted to supply the United States market. Allegations of corruption and human rights abuses have, however in the past blemished the relationship between Equatorial Guinea and countries such as the United States. The majority of the foreign oil companies in the country are from the United States, but China has grown in importance in Equatorial Guinea; CNOOC recently signed a production-sharing contract for an oil block in the country.

Concentration of Oil Supply in the Persian Gulf

Political risk is exacerbated by an increasing concentration of oil reserves in the Persian Gulf. Despite numerous recent successes, non-OPEC oil production will gradually be overtaken in the next two to three decades, leaving OPEC producers, particularly Middle Eastern producers, with increasing influence over the world oil market and the price of oil.

United States oil production has already begun its descent, with a 2005 output of 1.9 billion barrels of oil, compared to 2.7 billion in 1990 and 3.5 billion in 1970.[35] This is largely the result of the depletion of existing fields and the higher exploration and production costs of new reservoirs which are mainly located in challenging areas, including deep-water offshore and the harsh environment of Alaska.

Canada's oil production has been rising in recent years; oil production in 2004, averaging 3.1 million b/d, was a dramatic 7 per cent higher than in 2002. Canada's growth is the result of its new oil sands projects, which have more than compensated for the decline in conventional production. Commitments under the Kyoto Protocol and limitations on natural gas and water, both required for the extraction of bitumen, could however hinder further growth.[44]

Over the last twenty years, the growth of non-OPEC oil production has exceeded all expectations, but this trend cannot be extrapolated to the longer-term. Non-OPEC oil output will continue to increase throughout the decade before it reaches its limit, which is dictated by comparatively higher costs of investment and higher decline rates than those of the OPEC producers. The OPEC producers, therefore, will be in a much stronger position by 2020 as they begin to regain market share. The United States and Europe, along with other major oil importing countries, will have to contend with a new petroleum market with supply deriving overwhelmingly from the Persian Gulf. The IEA projects that OPEC's market share will increase from about 40 per cent in 2005 to close to 50 per cent by 2030.

Force Majeure

Force majeure, triggered by natural disasters, war or terrorism, can also significantly impact the security of supply on a global scale. This was underlined by the shutdown of production and refining in the Gulf of Mexico in the wake of Hurricanes Katrina and Rita, which made landfall during the Atlantic hurricane season of 2005. Hurricane Katrina, a Category 5 hurricane, demolished 46 platforms and damaged 20 others. During both storms, 100 per cent of Gulf oil production (about 1.5 million b/d) was shut-in and 94 per cent of gas production (about 280 million cubic meters per day) was shut-in during Hurricane Katrina. Although fears of supply shortage were deemed unfounded as the United States had commercial stockpiles to support demand for several weeks, oil prices nonetheless skyrocketed as a result of the shut down and the damage done to production and refinery facilities.

Terrorism aimed at pipelines, wells or other production facilities has plagued countries such as Iraq and Colombia, among many others, disrupting production and supply. In addition, petroleum companies in Nigeria, for example, have felt the intense pressure of civil disturbance in the Niger Delta for a number of years. As recently as January 2006, Royal Dutch Shell declared force majeure on 106,000 b/d of crude oil exports due to sabotage of the Trans-Ramos Pipeline. This event followed the kidnapping of four foreign subcontractors, who were later released after spending 19 days in captivity. As a result of these two incidents, 8 per cent of Nigeria's total output was shut-in. One year prior to this, ChevronTexaco stated that it was losing 140,000 b/d due to the suspension of activities in the Niger Delta. The company was also forced to stop production in March 2003 due to clashes between local ethnic groups. It is estimated that at the peak of the disturbances, 817,500 b/d, almost 40 per cent of Nigeria's production, was suspended.

While force majeure provisions in project agreements protect companies from contractual obligations and allow parties to avoid penalties for non-performance, the world markets remain vulnerable. In the immediate aftermath of Hurricane Katrina, gasoline prices in the United States jumped by 17 per cent to more than US$ 3 a gallon, while the price for gasoline, for example, in the United Kingdom rose 3 per cent from pre-Hurricane Katrina prices to US$ 7 per US gallon.

Infrastructure Bottlenecks

Massive investment in energy infrastructure will be required over the next few decades to ensure sufficient global supply. The IEA estimates a need for infrastructure investment of US$ 20 trillion (in constant 2005 dollars) from 2005 to 2030.[32] Nearly half of this investment will be required by developing countries as their demand and supply needs are projected to increase significantly. Though the majority of investment will take place in the power sector, investment in oil and gas infrastructure is estimated at about US$ 3 trillion. A regional breakdown of the energy investments required is provided in Figure 1.6.8.

Raising the necessary capital to finance these investments will be challenging. Since only a small percentage of the non-OECD governments are capable of funding these projects, private sector financing will prove crucial. Investment risks, however, are greater in emerging markets, and thus investors will require higher returns. Therefore, not only do the economic fundamentals of the individual projects have to be sound, but also governments need to establish favourable investment frameworks to attract adequate private sector involvement.

Figure 1.6.8 Energy Investment Required 2004-2030 (billion US dollars)

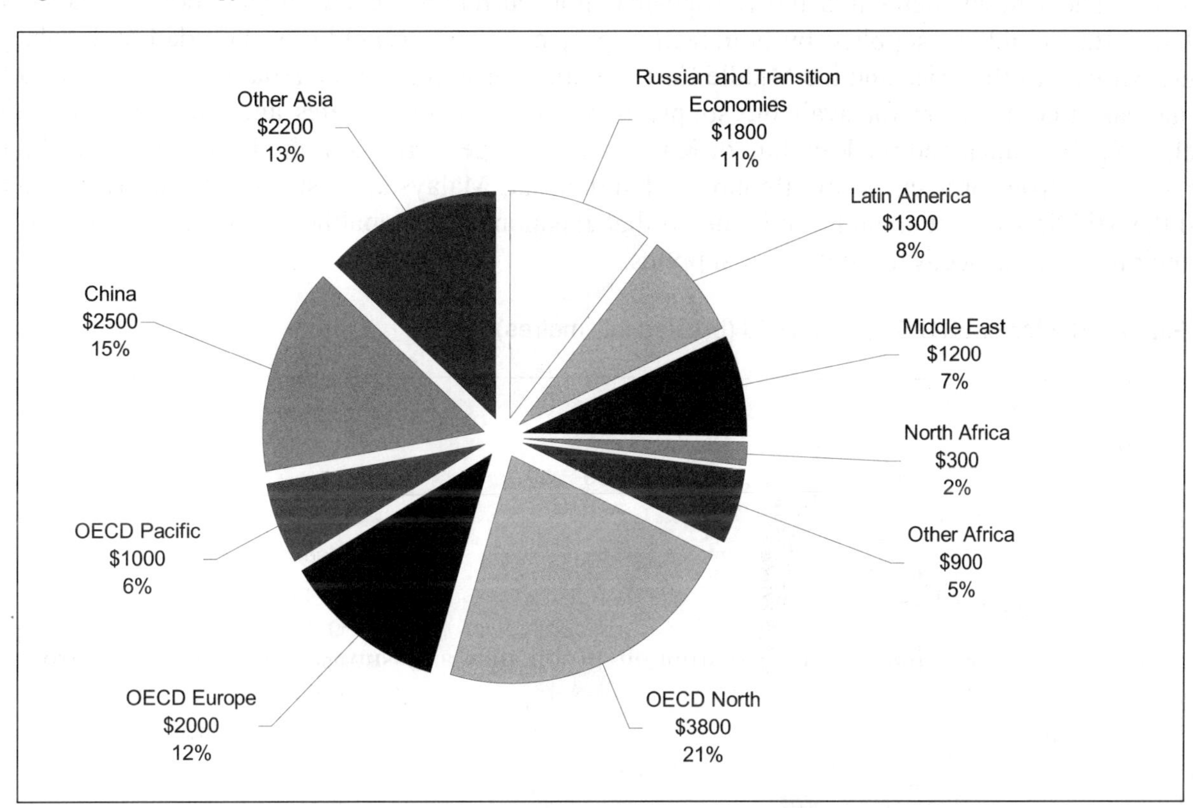

Source: IEA World Energy Outlook 2005

Deliverability – Focus on Liquefied Natural Gas (LNG)

As recent events have highlighted, one of the most exposed areas of energy supply is the import of natural gas by pipeline. Due to its physical nature, it can be difficult to diversify sources of natural gas supply. Once a pipeline is built, the buyer is locked in to that particular source of supply. Supply is then subject to all the risks that might potentially disrupt the operation of that single pipeline; risks such as natural disasters, political disagreements, operational accidents, etc. These risks can be diversified, to an extent, by building a second pipeline to an alternative source of supply. Geographic constraints, however, mean that many countries do not have the luxury of access to numerous sources of supply. In addition, building multiple pipelines is an expensive effort and may not always be economically feasible even when several supply sources are available.

One alternative that allows for better diversification of supply is LNG. LNG can be economically transported by specially designed ships to import terminals where the cargo is returned to its gaseous form and pumped into the domestic pipeline transportation network for distribution. From an energy security standpoint, an established LNG infrastructure provides flexibility that is not available with pipeline imports. In theory, supply disruptions at one, or even several, supply location(s) would not threaten energy security. Disruptions from one supplier can be made up by additional shipments from an alternate source, provided there was sufficient additional liquefaction capacity available. The same can be said of shipping and regasification facilities. If disruptions to shipping or import terminal operations occur, other ships can be found and shipments can be routed to alternate import locations. Flexibility in delivery infrastructure enables the risks to natural gas supply to be spread out over a variety of countries and facilities, allowing the benefits of diversification to reduce the overall risk.

The geography of the LNG trade divides the market into two major zones: the Atlantic and Pacific Basins. The United States and the European Union currently are the biggest consumers in the Atlantic Basin and are supplied by liquefaction projects in the Republic of Trinidad and Tobago, West Africa, North Africa and the Middle East. Supply constraints in the Atlantic Basin are leading to increased competition for available supply between the United States and European countries such as Spain. Japan and the Republic of Korea are the largest buyers in the Pacific Basin and they are supplied from projects in the Republic of Indonesia, Malaysia, Australia, Brunei Darussalam and the Middle East. It should also be noted that Japan and the Republic of Korea meet almost all of their natural gas needs through LNG supplies.

Figure 1.6.9 Major LNG Suppliers 2006 (billion cubic metres)

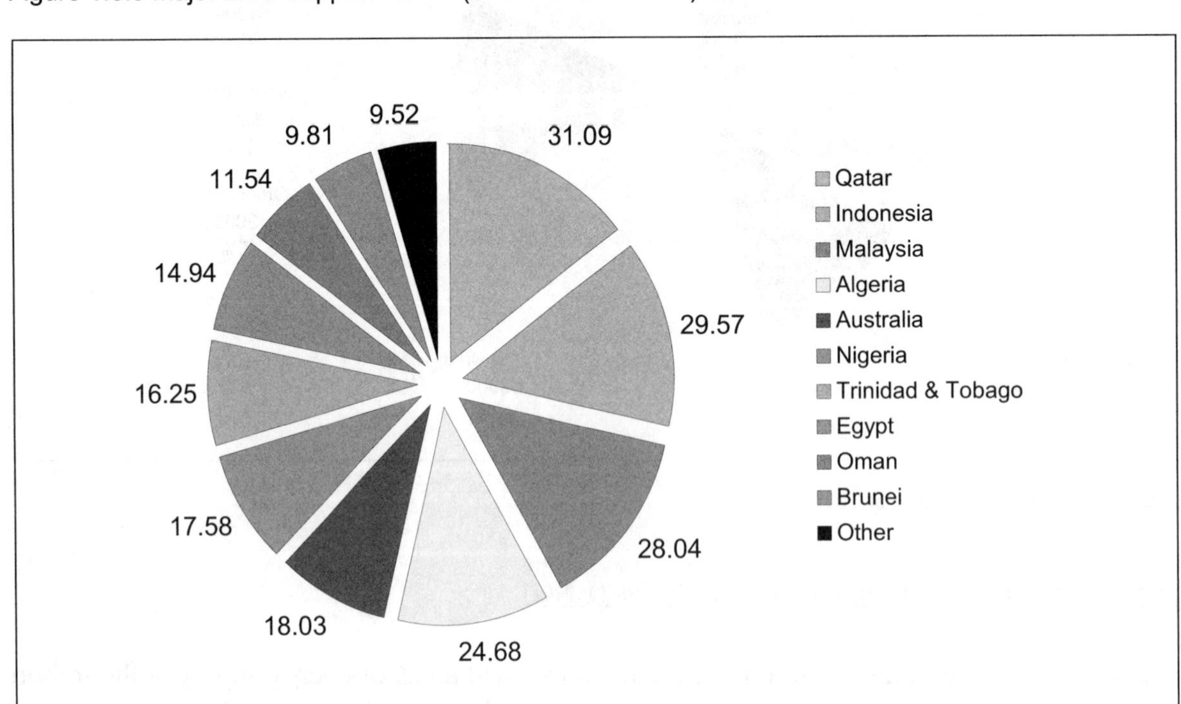

Source: BP Statistical Review of World Energy June 2007

Figure 1.6.10 Major LNG Buyers 2006 (billion cubic meters)

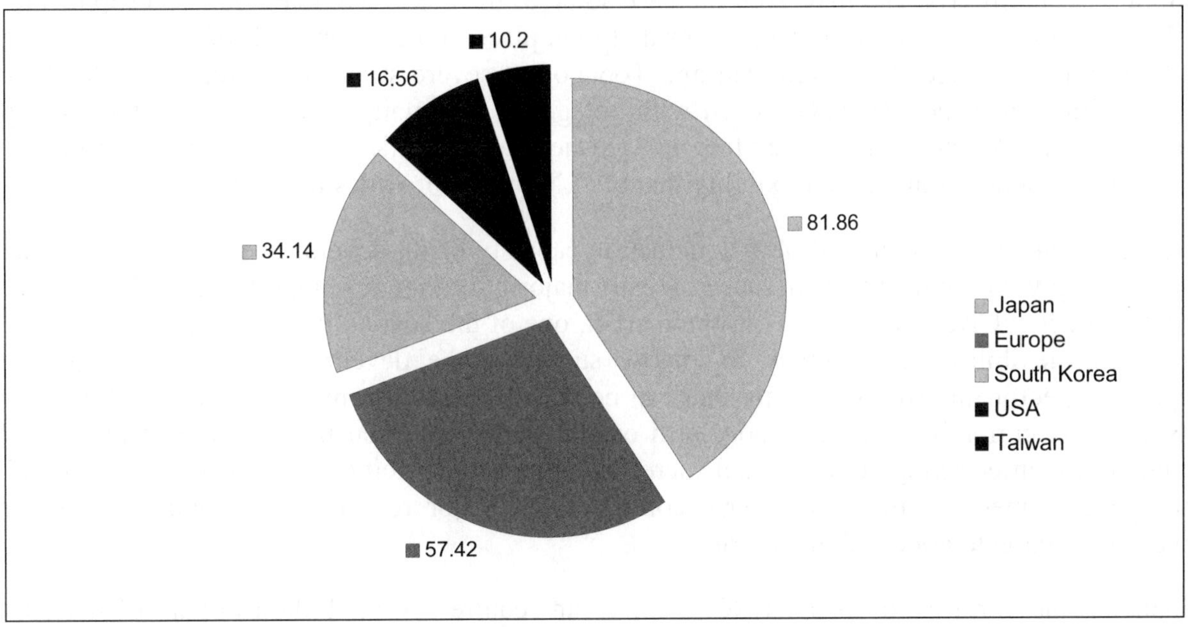

10.2
16.56
34.14
81.86
57.42

Japan
Europe
South Korea
USA
Taiwan

Source: BP Statistical Review of World Energy June 2007

Due to their geographic location, the countries of the Middle East are the only ones able to supply all three major markets (Asia, Europe and North America). Asian suppliers can economically ship cargos to the western coast of North America. It remains to be seen, however, if significant regasification capacity will be constructed in that region. Distance prevents LNG shipped from Asia to be sold economically into European markets.

Significant growth is taking place in the LNG trade. With regard to the Atlantic Basin, the State of Qatar and the Federal Republic of Nigeria are set to become the largest suppliers. Qatar will soon become the global leader in LNG liquefaction capacity and may reach over 70 mtpa (259 million cubic meters per day) by 2010. Nigeria is projected to have over 40 mtpa of capacity by 2010, and this figure could potentially increase. Algeria will continue to be an important supplier. Additionally, in 2005 Egypt began LNG production from two different projects. These two North African countries combined are now responsible for 40 mtpa of supply capacity. Angola, Equatorial Guinea, Norway, the Republic of Peru, the Russian Federation and the Republic of Yemen are all countries that could become new Atlantic Basin suppliers by 2010.

Destination flexibility is becoming increasingly important, especially with regard to trade in the Atlantic Basin. LNG buyers recognize the value of having regasification capacity in multiple locations on both sides of the Atlantic and taking advantage of arbitrage opportunities presented by price differentials. The same is true for suppliers that can also take advantage of price differentials. For example, supply from Trinidad can be sent to either the United States Gulf Coast or Spain, depending on where the price of natural gas is higher. This allows the market to respond to supply shortages or unexpectedly high demand.

Despite the anticipated growth previously mentioned, liquefaction capacity remains the greatest constraint to LNG trade in the Atlantic Basin. Liquefaction projects require billions of dollars worth of investment, and thus investments are only committed if price and demand can be guaranteed with a very high degree of certainty, meaning that close to all of the project's capacity is under long-term contract with off-takers. As a result, supply tends to track demand very closely, and there is limited excess capacity.

As it currently stands, the capacity constraint in liquefaction supply limits LNG's ability to reduce natural gas supply risk, and may even increase some of the risks. Tightness in LNG supply creates risks of its own, as can be seen from events that took place in the second half of 2005 when outages at LNG trains in Nigeria and Trinidad and Tobago put severe strain on producers in the Atlantic Basin to meet demand. However, even with outages at two major suppliers, the effect on supply was limited. Much of the quantities in question were made up by alternate producers or rescheduled, demonstrating the flexibility that the LNG trade provides to buyers.

Even with the large amounts of new liquefaction capacity being developed in the Middle East and Africa, supply will remain tight because the vast majority of that new supply is already under long-term contracts. Given the tightness in the market, one of the keys to increasing the role of LNG in providing flexibility and diversity to energy supply is the development of more liquefaction capacity, specifically excess capacity that can be sold on a short-term basis. The small amount of LNG that is currently sold on a short-term or spot basis has been made available mainly either through de-bottlenecking activities that increase nameplate capacity after a plant has been built or through customer default on long-term contracts. A major reason for this current situation is investor aversion to price and market risk.

Financing the construction of an LNG facility can require several billion dollars of investment. Uncertainty regarding the future price of and demand for natural gas means that investors will not provide capital unless they are assured of sufficient returns on their investment. This assurance is provided by selling almost all of a project's capacity under long-term sales contracts to creditworthy buyers before a project even begins construction. As a result, little capacity is left available for the more flexible short-term market.

Supporting the development of a short-term LNG market would make the LNG trade more flexible and would provide a cushion against supply shocks. Most LNG liquefaction projects provide the investment returns required by their investors at natural gas prices of US$ 3-4/million British thermal units (MMBtu). With prices at the United States' Henry Hub of US$ 8-9/MMBtu in 2005 and prices at the United Kingdom's National Balancing Point (NBP) even higher (see Figure 1.6.11), there seems to be an environment that should strongly encourage new development of and investment in LNG production capacity.

Figure 1.6.11 Natural Gas Prices ($/mmbtu)

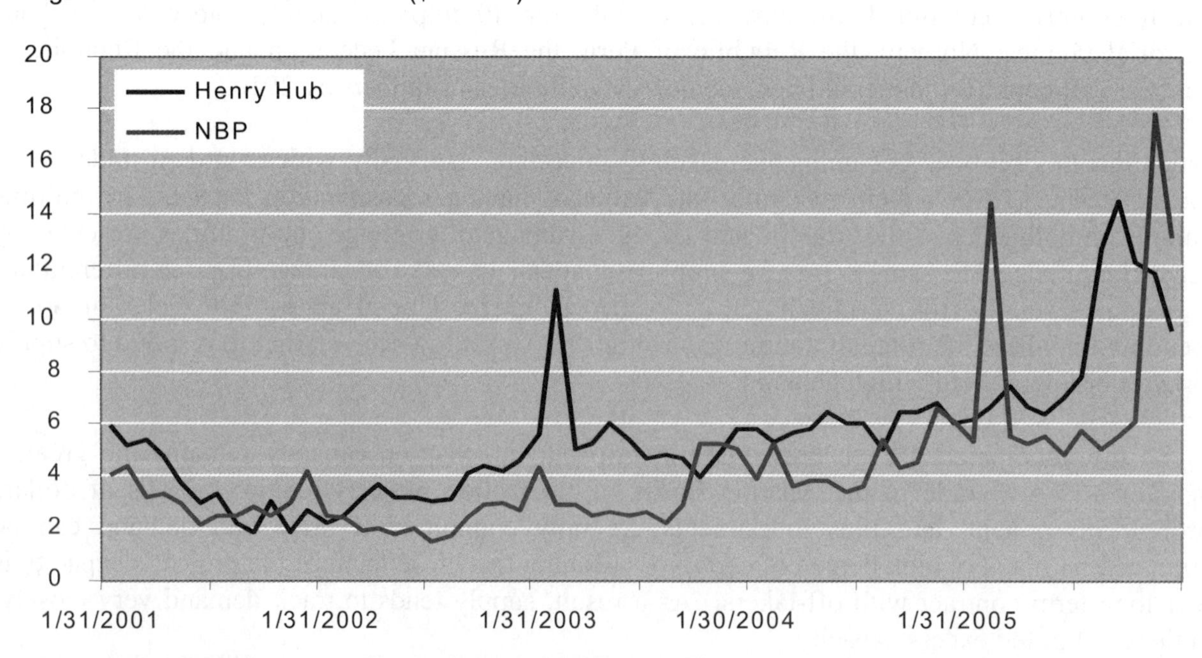

Source: United States Government, Energy Information Administration (EIA)

Currently the United States has five operating import terminals with a combined capacity of 25 mtpa (90 mcm/d) in the Atlantic Basin. Europe has twelve terminals in eight countries combining for over 50 mtpa (180 mcm/d) of capacity.[45] Expansions to existing terminals and greenfield developments are increasing the amount of re-gasification capacity available on both sides of the Atlantic.

United States policymakers have undertaken two major steps to encourage the construction of new LNG import terminals; (i) allowing new terminals to operate on a "closed access" instead of "open access" basis; and (ii) consolidating the authority for issuing permits within one regulatory body, the Federal Energy Regulatory Commission (FERC). Open access requirements would have allowed any company to bid on the use of the newly constructed import facility (as is the case for existing facilities in the United States); however, the FERC deemed that applying this regulation to new facilities would undermine investment in the needed infrastructure. By allowing closed access, also called "proprietary" access, investors are guaranteed access to the terminal capacity, reducing some of the uncertainty of the investment and enabling the project to be more easily financed. Secondly, by consolidating permitting authority within the FERC the regulatory process is streamlined and expedited. This arrangement means that developers only have to submit to a single body for regulatory approval, instead of two or three. The FERC still consults with state and local bodies on its decision, however, it holds the final word.

Despite these improvements in the regulatory environment, terminal developments in the United States have still faced a great deal of local opposition. The result of this opposition is that terminal projects have tended to cluster in areas such as the Gulf of Mexico, where the local population is more welcoming. A similar situation could be developing in the European Union whereby Spain and Italy may end up hosting the majority of the EU's LNG import capacity.

LNG can be a good option for diversifying natural gas supply and thus creating greater energy security. Nonetheless, LNG is not without its own risks. Due to current market dynamics, only a small amount of excess supply is available to respond to supply shocks. A large portion of the new import terminal development has occurred in geographic clusters, a tendency that reduces diversification benefits. Terminal development also bring about safety and environmental risks. However, almost all of these are risks, to a large degree, that can be directly managed by the importing country. This ability to be able to directly manage the possible risks to supply is perhaps the most important reason for encouraging further development of the LNG trade as a way to increase energy security.

The Role of Government in Promoting Energy Security

An energy security strategy should be centred on encouraging access to reliable resources at an affordable price. Security of supply can be enhanced by diversifying both the fuel mix and the sources of supply. Each, however, require substantial investments.

Producer governments do not need to provide the capital themselves for energy projects, but can achieve these objectives by ensuring social stability, creating a favourable investment framework and supporting research and development efforts to improve the recovery of their oil and gas reserves. Consuming governments should focus on improving and diversify transportation infrastructure (import as well as domestic), provide fiscal and legal conditions to encourage private investment in the energy sector and should consider special measures to foster investments in refineries, LNG terminals, cleaner coal technologies, nuclear and renewable energy sources in order to diversify their energy fuel mix.

Energy Exporting Countries

The governments of energy exporting countries need to ensure domestic social stability through the utilization of revenues for social, health and education investments. Having a long-term view on the utilization of oil revenues is crucial to ensure social stability. Oil revenues need to benefit as wide a portion of the population as possible. In this respect, the government's mission should be to set priorities for the use of oil revenues. This would help to combat what has come to be known as "the curse of oil" in many of the hydrocarbon producing countries in Africa and elsewhere.

Setting up an oil fund for use at the time when oil resources are depleted is one method of balancing short-term income with longer-term social investment. One of the best models of a successful oil fund was put in place by Norway. Net cash flows from oil activities are deposited into an account in the Bank of Norway and invested in long-term foreign securities. The fund functions as a national savings account to finance welfare programmes in provision for the day when oil resources are exhausted. In fact, part of the return on the fund's assets can already be used to finance Norway's social policies in case the budget is in deficit.

An oil fund is only effective, however, if the fund is managed efficiently and transparently. Nigeria's fund, established in 1999, has not been able to prevent social unrest. The World Bank estimates that over 50 per cent of Nigeria's oil revenues are earmarked by the Federal Government to service external debt and fund other projects, but the public perception is that oil wealth has not been appropriately utilized for social benefits. The Venezuelan Government's use of revenues from its national oil company, PDVSA, to fund social welfare programmes, on the other hand, have successfully reduced poverty and unemployment but have also constrained the possibility for investments by the company in the upstream and downstream sectors.

In addition to ensuring domestic stability, governments must ensure that national oil companies and private oil companies retain adequate revenues to finance new investments in production capacity. The problem is particularly acute for national oil companies, which in many countries have their budgets controlled by the state. This makes it difficult for companies to plan effectively for long-range development investments. In Mexico, for example, the effective taxation rate on Mexico's national oil company, Pemex, frequently equals or even sometimes exceeds 100 per cent.

Large capital investments are required at all stages of the oil and gas value chain: exploration, production, transportation, refining/ transformation and distribution. While the government can finance some of these investments, most of the capital will have to come from national or private companies, in particular international oil companies. In order to attract the required investments and develop the energy sector, governments must put in place a favourable investment framework, which allows national companies to retain sufficient funds for investment and allows private companies to invest at a profit.

Components of a favourable investment framework include:

(i) a stable political environment, with laws that protect investors against forced nationalizations or expropriations, and that guarantee equal rights for foreign and domestic investors (as most projects cannot rely on domestic capital alone);

(ii) a favourable tax regime. This does not necessary mean tax incentives for energy projects, although they are usually quite effective in attracting investors. Qatar, for instance, has been particularly successful in attracting investors by offering tax breaks for large capital investment projects in the LNG and petrochemical sectors. A country's tax regime should also be stable so that investors feel secure that they will not be exposed to unexpected changes in tax rates;

(iii) a stable currency to minimize foreign exchange risks;

(iv) well regulated capital markets;

(v) transparent permitting procedures and energy regulation, with an independent regulatory authority; and

(vi) a transparent judiciary system.

Several success stories illustrate the potential benefits of an attractive investment environment, improving the reliability of supply. Qatar is one such example. Over the past few years, Qatar has become a new entrant in international gas markets. Qatar's favourable reputation among international investors enabled the country to attract several billion dollars of foreign investment for its mega LNG projects of RasGas and Qatargas II. It has set a stable and reliable legal and regulatory framework that has made it possible to attract over US$ 40 billion of foreign investment in the energy sector over the past five years. Thanks to these investments, Qatar is poised to become the world's largest LNG producer after 2010.

Because of its stability, investors are less wary of investing major capital in the country as compared with more unstable countries elsewhere. Qatar has an A+ credit rating, a currency pegged to the US dollar, and generally sound economic policies. The country's constitution offers strong protection of foreign investor's rights against expropriation and guarantees equal treatment. In addition, to attract more foreign capital, Qatar has eliminated most restrictions on FDI and offers fiscal incentives in the oil and gas sector. The Qatargas II Project received an 8-year exemption of income tax for LNG destined to the United States and the United Kingdom. The tax regime was important in securing the total investment for such a large transaction.

Trinidad and Tobago is another example of a small country taking active steps to attract foreign direct investment and converting its available natural resources into jobs, social stability and economic prosperity. Trinidad and Tobago's political stability and well-educated workforce are important components of its favourable investment climate. Trinidad's foreign investment law opens the country to foreign investment in almost all sectors and guarantees equal rights to foreign investors. The country also has a developed and well-regulated financial sector. Taxation is transparent and relatively moderate, with a corporate income tax capped at 30 per cent. The Government of Trinidad and Tobago is also taking a proactive approach to foreign investment, openly inviting foreign investors to invest in the country's energy sector. Foreign direct investment is expected to play a major role in the development of Trinidad's petrochemical sector.

Both Qatar and Trinidad and Tobago serve as examples to other producing countries seeking to attract capital for large energy infrastructure projects.

To enhance the recoverability of oil and gas reserves, governments of producing countries, together with international organizations, can sponsor research and development initiatives to improve the exploitation of oil and gas reserves. New tertiary techniques, as well as the development of unconventional sources of oil and gas, can contribute to increasing the availability of oil and gas resources previously considered unrecoverable.

Energy Consuming Countries

Governments of energy importing countries have a role to play in improving their energy security; but it is not so much in providing capital as in providing a favourable environment so that investments in vital energy infrastructure can take place.

Besides policies aimed at diversifying energy sources and improving energy efficiency, governments can also play a role in the construction of transportation infrastructure, such as pipelines. The European Union has chosen to actively intervene, while the United States has pursued a more laissez-faire approach to energy transportation infrastructure. The United States has an extensive pipeline network, both within the country to service all consuming regions and to connect the country with Canada and Mexico. The total cumulative length of oil and natural gas pipelines on United States territory is about 500,000 miles. All pipelines are privately operated. There is no national pipeline strategy contrary to Western Europe. FERC, a federal government agency, provides the regulatory supervision for the permitting of construction of pipelines and also supervising safety. Market forces largely determine which pipelines are built and which are not. Despite this, there is strong redundancy and therefore flexibility in the system.

The European Union has been more active in developing an energy transportation strategy known as the Trans-European Networks strategy, which puts particular emphasis on pipelines. The initiative launched in 1994 helps fund priority projects. During the period 1994 to 2003, priority was given to interconnections between EU countries, so that countries could help each other in the event of a supply disruption. Before 1994, pipelines were the responsibility of individual member states and followed mostly national strategy. As a result, few cross-border pipelines existed and networks were poorly connected. The Trans-European Networks strategy has helped correct this, but energy distribution still remains largely national in scope. For the period 2003-2010, priority has been given to diversifying sources of energy supply and better connecting the EU with its major suppliers in North Africa and Russia.

Consuming governments can help to enhance energy security by facilitating the construction of LNG import terminals. The United States and Spain have taken different approaches to their respective regulation of new LNG import terminals. New terminals in the United States operate in a comparatively deregulated environment. Terminal owners are allowed to set their own tariffs and operate their facilities on a closed access basis. This allows investors to be secure in the knowledge that they will have access to their facilities and earn a return sufficient to make the investment worthwhile. As such, it provides a stable foundation for terminal development. The downside is that there is no guarantee that terminal owners will allow third party access to their terminals. Competition may be limited to supply from rival terminals, as opposed to competition among rival suppliers at a single terminal. Lack of open access also means that some terminals may not be open to short-term or spot deliveries.

The Spanish model is much more regulated. The Spanish regulator bases the regulated return on a 75 per cent load factor and additionally limits the proportion of terminal capacity that can be under long-term contract to 75 per cent. Through this approach, Spain has ensured that if terminal owners can secure long-term contracts for a large portion of terminal capacity then they can be sure of a reasonable return while also guaranteeing that there will be some excess capacity in the system for short-term LNG trade use. The United States will also almost certainly have significant excess terminal capacity, but there is no corresponding guarantee that this capacity will be available for the more flexible short-term market.

Due to the current insufficient global oil refinery capacity, consuming governments should take steps to encourage investment in domestic refining capacity in order to enhance flexibility and to improve supply security of oil products. The current trend is for refinery capacity to be constructed close to crude oil supplies, namely in oil producing countries, which serves to exacerbate the risk in the event of a crude oil supply disruption. Governments of consuming countries, such as the United States, should consider streamlining their permitting processes to reduce the development phase of refinery investments. In addition, consideration should be given to provide inducements for new capacity additions. This would lower the cost of potential projects and help ensure that refineries are constructed in a timely manner and closer to consuming centres.

Governments have a special responsibility when it comes to the development of new and alternative energy technologies to deal with energy-related environmental problems or other social objectives, such as, improving energy security. One such area is clean coal technologies. Currently, the private sector is unable to assume the financial risks of investing in untried low-emission, low-cost clean coal technologies, and therefore projects can only proceed as public-private partnerships (PPP), with strong government, university, and industrial leadership. The United States Energy Policy Act of 2005 and the Transportation Bill provide significant federal financial incentives for a variety of such coal gasification projects. But other countries need to consider similar measures in order to support the commercialization of a range of clean coal technologies.
Likewise, investments in nuclear power and renewable sources of energy require strong government support to make them economically viable. Governments can foster development of these sources by providing direct subsidies, tax credits, and/or fixed off-take prices. By doing so, consumer countries can diversify their energy sources and reduce their dependence on fossil fuel imports.

With appropriate government support energy security through diversification of supply is attainable. However, large investments in supply and transportation infrastructure are needed to make this diversification a reality – investments that could reach into the hundreds of billions or even trillions of United States dollars. The public sector does not have the resources to make these investments alone. However, strategic policies can encourage the participation of the private sector. By providing the right combination of stability and incentives, governments can mobilize the resources of the marketplace to build out the required infrastructure and make the investments in technology necessary to provide options and flexibility in supply. Increased supply options and flexibility reduce risk by enabling a rapid response to shocks, thus increasing energy security.

2.1 REPORT ON GLOBAL ENERGY SECURITY AND THE CASPIAN SEA REGION

Introduction

Energy security has re-emerged as a crucial issue for policymakers, energy industries, the financial community and the general public. There is an increased sense of unease concerning secure access to reliable supplies of reasonably priced energy.

Concern about energy security in UNECE countries has grown and waned over the years. This problem was uppermost in the minds of energy policymakers during the 1970s and early 1980s when energy supply and demand were tightly balanced and energy markets were rocked by two sharp oil price rises. Likewise, concerns were renewed and heightened during the Iraqi-Kuwaiti crisis of 1991.

Many factors have contributed to this heightened sense of vulnerability: the recent high crude oil prices and their volatility; the increased cost of developing incremental sources of supply; longer supply routes; the instability in Iraq; the tensions in the Middle East; sabotage and terrorist attacks in major oil and gas producing countries; the unfavourable investment climate in a number of producing countries; the growing energy import dependence of major consuming countries; and the corporate and policy failures, such as the Enron bankruptcy and the 2003 electric power blackouts in North America and Europe.

The objective of this section is to examine the potential contribution of Caspian Sea countries, in respect to oil and natural gas and, to some extent coal, in mitigating global energy security risks. In order to provide a context for this assessment, the section begins with a brief review of the objectives of energy policy. Then energy supply and demand trends, including the contribution of Caspian Sea countries to future fossil fuel supplies, and their consequences for energy security are examined. This is followed by a brief discussion of the implications of terrorism on energy security. The report finishes with a brief overview of conclusions and recommendations regarding developments in the Caspian Sea countries and their potential contribution in improving the energy security of UNECE countries.

Energy Policy Objectives

A sustainable energy development strategy for the UNECE region must have at its core the following policy objectives:

(a) secure access to high quality reasonably priced energy services for all individuals in the UNECE region in the short, medium and long-term;

(b) reduction in health and environmental impacts resulting from the production, transport and use of energy;

(c) well-balanced energy network systems across the whole of the UNECE region tailored to optimize operating efficiencies and overall cooperation among countries;

(d) sustained improvement in energy efficiency, in production, transport and use of energy; and

(e) steady reduction in energy-related health and environmental impacts through the development and application of environmentally sound and economically viable technologies.

In many respects, energy security is the raison d'être for energy policy. Energy and energy industries are vital to all modern societies. They underpin economic growth and development. They contribute to the material well-being and comfort levels of populations and they touch every aspect of the daily life of individuals. As a result, governments have historically felt the need to pay special attention to energy and particularly energy security.

While energy security is a multifaceted concept and not easy to define, there are four dimensions of particular relevance: physical disruption of supplies due to infrastructure breakdown, natural disasters, social unrest, political action or acts of terrorism; long-term physical availability of energy supplies to meet growing demand in the future; deleterious effects on economic activity and peoples due to energy shortages, widely fluctuating prices or price shocks; and collateral damage from acts of terrorism resulting in human causalities, serious health consequences or extensive property damage.

Energy Demand and Supply Trends

The energy marketplace is in a constant state of flux and change. There are numerous trends and developments in the regional and global marketplace that can affect the energy security of countries in the UNECE region. These include the future evolution of energy demand, the sources of future supplies to satisfy this demand, the variety and diversity of fuels and energies that might be available to consumers in the years ahead, the geographical distribution and concentration of fossil fuel production and reserves, the potential use of market power, the diversity and reliability of energy transportation infrastructure, and the level of social unrest and ethnic strife in producing and transit countries.

The energy import dependence of many UNECE countries will continue to rise for the foreseeable future. In most countries, the growth in energy demand, buoyed by growth in transport and electricity use, is expected to outpace the growth in domestic energy production. The major exceptions are the fossil fuel rich UNECE countries, such as the Russian Federation, Norway, Canada, Azerbaijan and Kazakhstan, which will remain major producers and exporters of fossil fuels.

Oil

The increased sense of vulnerability and insecurity with respect to oil is fuelled by concerns regarding: the growing dependence of UNECE countries on imported oil; the concentration of known oil reserves in the Middle East; the growing dependence on the Middle East and, more generally, on OPEC countries; the perceived higher costs of new incremental oil supplies; the ever increasing distance of supply routes connecting producing and consuming centres; and the potential for political instability and social unrest in some of the major oil producing and transit countries.

The dependence of countries in Western Europe on oil imports, which today stands at approximately 60 per cent, is likely to rise to about 80 per cent by 2030. In North America, import dependence could rise from about 40 per cent to 50 per cent by 2030. In the case of Central and East European countries (excluding the Russian Federation), oil import dependence, which is currently more than 80 per cent, could rise to well above 90 per cent by 2030.[46] Hence, in the absence of measures to offset increased oil import dependence, UNECE countries could become more susceptible to world oil supply disruptions or other shocks over time.

Currently, the Middle East supplies about 30 per cent of all the oil consumed in the world. By 2030, this could be up to 40 per cent.[46] Production costs are among the lowest in the world and so are investment costs. Moreover, approximately two-thirds of the world's established reserves of crude oil are in the Middle East. With time, reliance on the region for oil is bound to rise, assuming there is no widespread introduction of alternative fuels for transport or reduced demand flowing from more efficient vehicle technologies (e.g. hybrid and plug-in hybrid vehicles).

Similarly, the world's dependence on oil from OPEC will continue to rise over time. Today, OPEC's share of world oil production is around 40 per cent. It is likely to rise to about 50 per cent by 2030.[46] This is very close to OPEC's market share of 54 per cent reached in 1973 at the time of the first oil crises.

The Islamic Republic of Iran, bordering the Caspian Sea as well as the Persian Gulf, is an important member and the second largest oil producer in OPEC. The crude oil potential of the country is large. The Islamic Republic of Iran currently produces about 4.3 million b/d[47], which represents over 5 per cent of total world crude oil production. With proved oil reserves of 138 billion barrels[47], about 11.5 per cent of the world's total, the Islamic Republic of Iran is likely to be able to significantly increase its oil production in the future as it invests in both existing and new fields, including the Caspian offshore.

Likewise, crude oil production from the Russian Federation and the other Caspian Sea countries is expected to rise over the foreseeable future. The Russian Federation in particular has experienced a significant increase in production in recent years. Today, with oil production of 9.8 million b/d[47], the Russian Federation accounts for approximately 12 per cent of total world production. This level is expected to rise to approximately 11 million b/d over the medium-term. While much of this production is expected to come from existing fields, new greenfield projects are expected to be developed in the Caspian Sea, East Siberia and Sakhalin.

Oil production from the other Caspian Sea countries, that is Azerbaijan, Kazakhstan, Turkmenistan and Uzbekistan, is about 2.4 million b/d[47], accounting for about 3.0 per cent of total world production. Proved reserves and resources are substantial, particularly in Kazakhstan and Azerbaijan. With proved reserves of about 46 billion barrels[47] in Kazakhstan and Azerbaijan combined, production is bound to rise. In addition, while proved oil reserves in Turkmenistan are small, the ultimate recoverable resources are estimated to be significant. Therefore, most projections call for oil production from the Caspian Sea region countries of Azerbaijan, Kazakhstan, Turkmenistan and Uzbekistan to rise over time and possibly double over the next 10 years.

Most oil exports from the Caspian Sea region, including those intended for western European customers, have so far transited through the Russian Federation to market. Adequate oil transport facilities from the region have in recent years been a constraint on exports.

In order to accommodate increased exports from this region, existing transport routes through the Russian Federation have been expanded. Similarly, new alternative routes are being planned and developed. In that regard, the Baku-Tbilisi-Ceyhan crude oil pipeline (from Azerbaijan, through Georgia and across Turkey to the Mediterranean coast) began operation in the fourth quarter of 2005.

Another existing transportation constraint for oil exported from the Caspian Sea region to Europe is shipping congestion through the Bosphorus and Dardanelles straits that separates the European and Asian parts of Turkey. About 3 million b/d of oil, including some oil originating from the Caspian Sea region, transits through the straits from the Black Sea to the Mediterranean. Because the Bosphorus and Dardanelles straits are relatively narrow, the Turkish Government introduced a Vessel Traffic and Management System to assist in the passage of tankers through the straits for safety and environmental reasons.

Taking into account the current congestion of tanker traffic through the Bosphorus and Dardanelles straits and the potential for significant increased exports from the Caspian Sea region, alternate routes that bypass the straits will have to be developed for safety and environmental reasons if maritime oil transport in the Black Sea is to grow.

In addition to exports to the west, increasing attention is being given to the potential of exporting oil eastward from the Russian Federation to China and Japan and from Kazakhstan and Turkmenistan to China. Asian demand for oil is growing very rapidly, particularly in China and India. The rapid pace of industrialization and economic development in both countries has meant a high rate of consumption of energy and oil. The oil producers on the eastern side of the Caspian Sea are well positioned to take advantage of market opportunities in Asian countries if transportation routes, whether it be by pipeline or rail, can be economically developed.

Natural Gas

Natural gas is today's fuel of choice. It is flexible to use, environmentally friendly compared to other fossil fuels, relatively abundant, with supplies perceived to be relatively secure and reliable. Consequently, it is being used in a variety of sectors and applications, and is experiencing significant growth as a fuel for electricity generation.

However, the rapid growth in natural gas consumption is boosting the import dependence of many European countries. While this may not be a problem in the short to medium-term, meeting demand over the longer-term could become a challenge as new sources of supply become increasingly more remote and more costly to develop. Likewise, demand for natural gas is rising in the United States and its dependency on foreign sources will continue to grow, with the shortfalls increasingly being meet with imported LNG.

The import dependency of Western European countries is set to rise from about 40 per cent of natural gas consumption to over 50 per cent by 2015 and continue to increase thereafter even assuming a significant expansion in Norwegian production. The import dependence of Central and Eastern European countries, excluding the Russian Federation, is also likely to rise from about 65 per cent to 85 per cent by 2015.[46]

On the other hand, the situation in North America is somewhat more encouraging; up to now the market has been relatively self-sufficient, with gas supplies and transportation infrastructure well balanced and diversified. However, there are signs that increased LNG imports will be required over the medium to longer term. In order to accommodate these rising import needs, additional LNG gasification plants and export terminals will be required in producing countries and new re-gasification infrastructure built in the United States. It is highly unlikely that imported LNG into the United States over the medium term will be sourced from natural gas produced in the Caspian Sea region. However, over the longer-term some could be based on natural gas from the Caspian Sea region, liquefied on the Baltic Sea coastline of the Russian Federation or along the Mediterranean in Turkey.
The problem of import dependence is compounded when countries have to rely on a single outside source of gas. Most countries in Western Europe are now supplied from a number of sources, including indigenous sources of supply. For historical and geographical reasons, this is not generally the case for countries in Central and Eastern Europe. Almost all the gas imported in these countries, to supplement domestically produced gas, comes from the Russian Federation.

To date, the Russian Federation has been a secure and reliable supplier of natural gas to both Central and Eastern as well as Western European countries. Since deliveries began over 35 years ago there has been no major interruption of gas supplies. However, the recent dispute (at the beginning of 2006) between Gazprom (the Russian Federation) and Naftogaz (Ukraine) over natural gas prices for imported Russian gas and transit fees, and the ensuing short-term disruption of natural gas supplies transiting through the Ukraine to Central and Western Europe, did raise concerns about security of natural gas supplies in downstream importing countries.

The traditional suppliers, like the Russian Federation, Algeria, Netherlands and Norway, are likely to have the capacity to meet Europe's growing demand for natural gas for some time to come. However, in the longer-term, significant new investments in production and transportation infrastructure will be required. Moreover, supplies will increasingly have to be transported over longer distances as new production centres are developed in more remote or distant areas of the Russian Federation, the Norwegian shelf, North Africa and the Caspian Sea, and which may ultimately also include supplies from the Islamic Republic of Iran.

The Russian Federation is the largest producer of natural gas, accounting for about 21 per cent[47] of total world gas production. It is also the country with the largest proved gas reserves, 48 tcm[47] or 26 per cent of the world total. While new incremental gas supplies are likely to come from smaller fields, from more remote and harsher environments and from geologically more complex structures, Russian natural gas production is expected to expand from about 600 to 750 bcm over the next 10 to 15 years. Likewise, exports are projected to rise to Western Europe and to Eastern Europe and could perhaps even begin to Asia over the next 15 to 20 years.

The Islamic Republic of Iran is a major producer of natural gas, but it is not yet a major exporter. However, its proved reserves are large compared to its production. Therefore, it has the potential to significantly increase production if economically viable markets can be developed. At the moment, there are some quantities of gas exported to Turkey and some gas imported from Turkmenistan, but the quantities involved are small relative to the resource base.

While proved natural gas reserves in Azerbaijan, Kazakhstan, Turkmenistan and Uzbekistan are not of the same magnitude as those in the Russian Federation and the Islamic Republic of Iran, they are nonetheless substantial. The four countries combined have 9 tcm of proved reserves, representing about 5 per cent of total world reserves. [47] Ultimate recoverable resources are however considered to be significantly higher.

Currently, most of the gas produced, about 149 bcm[47] per year, in Azerbaijan, Kazakhstan, Turkmenistan and Uzbekistan is used mainly for internal domestic consumption. However, there are exports to the Russian Federation and beyond, such as to the Ukraine, through the existing Russian network. Gas production and exports from the four countries are expected to grow over time. Gazprom has already contracted to buy 1 tcm out of the 9 tcm of proved reserves under long-term contracts.

In addition to exports through the Russian network, a gas pipeline is under construction from Azerbaijan to Turkey through Georgia (the South Caucasus Pipeline Project/ Baku-Tbilisi-Erzurum pipeline) that will increase export capacity from the region. There are also other projects under consideration, such as the Nabucco Project (Turkey-Bulgaria-Romania-Hungary-Austria line) and the Turkey-Greece gas pipeline interconnector, which, if constructed, would provide reliable and secure transit routes for natural gas from the Caspian Sea to markets in Southeast and Central Europe.

In the longer-term, additional export capacity could be built to supply Western Europe and ultimately to supply China and other Asian markets, in the latter case particularly from Turkmenistan. But this will depend on success producers are in proving up additional reserves and on the economic and geopolitical situation in the region.

The transit of natural gas through the territory of third countries continues to be an issue of controversy and potential tension. Transit rights, which are of concern to both gas-exporting and gas-importing countries, are sometimes the subject of intense commercial and political negotiations. The potential for disputes and misunderstandings are ever present in a number of regions of the UNECE, including the Caspian Sea region and the Caucasus.

Coal

From the point of view of energy security, coal has the advantage that global coal reserves are large; sources of supplies are diversified; ample supplies are available from politically stable regions; world infrastructure is well developed; new supplies can be easily brought on stream; and coal can be stored.

The Russian Federation, which has 17 per cent of the world's proved reserves, and Kazakhstan, which has 3.5 per cent of the global total, are major coal producing countries.[47] In the case of the Russian Federation, most of the coal produced is consumed domestically. On the other hand, in the case of Kazakhstan, about 40 per cent is exported. In addition to being able to use an indigenous energy source domestically, the added benefit to both the Russian Federation and Kazakhstan is that coal can displace the use of oil and natural gas in the domestic market, thereby, making it possible to export more oil and natural gas and, thus, contributing to higher export revenues.

While coal is plentiful and secure, and its greater use both in the Russian Federation and Kazakhstan could release more oil and natural gas for export, coal is besieged with environmental problems. These problems can be mitigated to some extent by the use of existing, commercially viable clean coal technologies. However, for coal to completely shed its negative environmental image, the commercialization and wider use of new technologies (involving gasification, liquefaction and carbon capture and storage) will be required.

It is imperative for governments and the private sector to invest in research, development and the deployment of new technologies that reduce the health and environmental impacts of fossil fuels – oil, natural gas and coal. While improvements in energy conservation and efficiency can slow down the rate of growth of demand for energy services, they cannot eliminate this demand altogether. Energy services are and will continue to be needed to meet human needs and for economic development. Under present and projected market conditions, the bulk of these energy services will be provided by fossil fuels.

Projections by a range of organizations indicate that oil, natural gas and coal are likely to remain the mainstays of global energy supply systems for the greater part of the twenty-first century. Consequently, it is important that the energy systems and energy infrastructure, including the oil, natural gas and coal resource base, of the Caspian Sea region be interconnected and integrated into the European and world energy systems.

Acts of Terrorism

No region of the world is immune to sabotage and acts of terrorism, including the Caspian Sea and Caucasus regions, which continue to be the subject of social, economic and political tensions.

Following the tragedy of 11 September 2001 in New York and Washington, the world entered into a new era of insecurity. The events of this day ushered in a new kind of international security risk; what used to be figments of an imaginative mind – the use of civilian infrastructure and commercial airliners as weapons by terrorists to cause mass destruction – became real and genuine.

The events of 11 September also raised a number of questions about the vulnerability of energy infrastructure to terrorist attack. Energy infrastructure is composed of primary energy-producing units that are connected to energy markets through a vast system of pipelines, road and water transportation, and electric power grids, including storage facilities. This complex, vast and expensive infrastructure is an appealing target for sabotage and acts of terrorism.

Even prior to 11 September, there were instances of acts of sabotage and terrorism on oil and gas pipelines and high-tension electricity transmission infrastructure. For the most part, however, these acts, aimed at furthering economic, political or ideological objectives, were usually localized and intended to disrupt the flow of energy and create economic hardship. It is clear that 11 September changed all of this. Today, one cannot dismiss the possibility of acts of terrorism on energy infrastructure aimed at generating the most extensive collateral damage possible, such as loss of life and mass destruction of property.

Large oil and gas production facilities (e.g. offshore platforms) and storage facilities (e.g. large liquefied natural gas storage tanks located in or near urban centres) have the greatest potential for creating collateral damage, though they may not necessarily be the most susceptible to attack. Likewise, attacks against thermal power plants and associated facilities, particularly in urban areas, can cause significant collateral damage.

On the other hand, oil and gas pipelines, and associated compressor stations, while more susceptible to sabotage and terrorism, are much less likely to cause widespread loss of life and property damage. Terrorist acts against these facilities, as well as electricity transmission infrastructure, can disrupt the availability and normal flow of energy, and if perpetrated against transportation hubs, can compromise the flexibility and integrity of transportation networks. However, they are less likely to result in significant collateral damage and can be repaired relatively quickly.

There are a variety of risks associated with nuclear power ranging from theft, sabotage or illicit trafficking in nuclear material and other radioactive substances, to sabotage or acts of terrorism against nuclear facilities or transport systems. There has been substantial international cooperation both to upgrade facilities around the world and to put more effective security recommendations and standards in place. Peaceful nuclear installations are often more robust and much better protected than other hazardous non-nuclear installations.

The security risks and consequences of an attack by terrorists on energy installations need to be assessed and taken into account in energy policy decision-making. Facilities and hazardous materials need to be protected from mass-consequence sabotage or acts of terror. Adding to the woes of the energy sector are the difficulty and expense of obtaining insurance coverage against terrorism, hence exposing energy companies to potentially large financial risks.

It should though be stressed that while the potential for acts of terrorism are real and ever present, measures taken by governments and the private sector to protect energy infrastructure have to be commensurate with risks and potential damage. In the final analysis, these measures will invariably have to be paid for, and in most cases this will be by energy consumers.

Conclusions

Hydrocarbon production and exports from the Caspian Sea region will significantly increase over the next ten to fifteen years. This will contribute to meeting growing global energy demand as well as helping to mitigate, to an extent, global energy security risks. However, it is important to keep in perspective that, while hydrocarbon production from the Caspian Sea region will increase substantially and contribute much to the economic development of the region, its overall contribution to meeting global demand for hydrocarbons will still remain modest. The region currently accounts for about 2 per cent of total world hydrocarbon production and it is not likely to contribute much more than 3 per cent or so of total world production over the medium term.

To accommodate the growing export of hydrocarbons from the region, existing export transport capacity has been expanded, new transport corridors developed and new pipelines constructed. Other export transport routes are under development or consideration. Most existing and new export transport routes are oriented westward to supply western industrialized countries, although some capacity is now being developed to supply Asian countries, notably China. At this time, the total transport capacity, existing, under development and planned, would seem to be adequate to meet expected export requirements.

The Bosphorus and Dardanelles straits, linking the Black Sea with the Mediterranean, are already well congested with oil tanker traffic. About 3 million b/d of oil, some of it Caspian Sea oil, transits from the Black Sea to the Mediterranean through the Bosphorus and Dardanelles. A major accident could cause significant environmental damage. Therefore, Caspian Sea oil exports, together with oil from other sources, will increasingly need to bypass the Bosphorus and Dardanelles either by way of pipelines through Turkey and the Russian Federation and/or across the Black Sea to ports in Bulgaria, Romania and the Ukraine.

The development of further oil and natural gas transport infrastructure linking the Caspian Sea region to Southeast and Central Europe would contribute to diversifying the hydrocarbon sources of supply of the countries in those two regions, enhance security of supply and provide economic development opportunities. Countries along transport routes would also benefit from transit fees. This intensified cross-border infrastructure would in addition promote greater mutual interdependency among countries, enhance regional integration and, thereby, contribute to greater economic and social stability.

A legal framework for the Caspian Sea is required. While not completely constraining development of the Caspian offshore, the lack of a legal framework is nonetheless hampering the planning and development of some offshore fields where conflicting territorial claims exist. Likewise, cooperation among the littoral Caspian Sea countries on environmental matters, including biodiversity, is essential in order to protect the marine environment for future generations as resource development proceeds in the Caspian Sea.

While Caspian Sea hydrocarbon resources are important to the Russian Federation and the Islamic Republic of Iran, both countries have access to significant hydrocarbon resources elsewhere on their territories. As such, they have a wide portfolio or wide range of options regarding hydrocarbon development projects. The same, however, does not apply to Azerbaijan, Kazakhstan and Turkmenistan where hydrocarbon resources are largely concentrated in or adjacent to the Caspian Sea. Hence, the development of these resources and issues relating to export outlets are much more important to these three countries than to the Russian Federation or the Islamic Republic of Iran. This is because of the more limited range of development opportunities and the contribution that these resources can make to the respective economies of the three countries in terms of economic activity, employment and fiscal and export revenues.

Political differences and tensions among countries within and outside the Caspian Sea region continue to hamper cooperation among the countries for energy development and use. Likewise, inadequate economic development opportunities and social and ethnic unrest in neighbouring countries continue to trouble the region as a whole, undermining the opportunity to build a prosperous and secure future. If political, economic and social problems could be overcome, an integrated hydrocarbon production, transportation and utilization infrastructure network could be developed extending from western Europe in the west to India in the east.

International cooperation and producer-consumer dialogue are key elements for establishing the investment environment and regulatory framework that Caspian Sea region countries will need in order to attract the investments required to further develop their hydrocarbon reserves and resources and to extend their energy transportation infrastructure more widely.

Recommendations

The following recommendations on enhanced trade and international cooperation for the littoral countries of the Caspian Sea emerge from the preceding analysis and the deliberations of the UNECE Energy Security Forum High-Level Meeting on the Caspian Sea Region (28 June 2005) and the Seminar on Energy Security Risk Mitigation in the Caspian Sea Region (21 April 2006).

To enhance energy trade and cooperation, the countries of the Caspian Sea region should:

(a) Cooperate on the analysis of the geological conditions, measurement and assessment of hydrocarbon deposits to more accurately determine the size, scope and nature of the energy reserves and resources in the region;

(b) Conclude a legal framework for the Caspian Sea in order to eliminate conflicting territorial claims and ensure the protection of the marine environment for future generations as resource developments proceed in and adjacent to the Caspian Sea;

(c) Appraise the geological conditions of proposed hydrocarbon pipeline corridors and the consequences of seaborne oil transport to ensure the public safety and environmental integrity of the region;

(d) Establish and maintain clear, transparent and unambiguous legal requirements and regulations on investments in order to develop and transport hydrocarbon resources to international markets;

(e) Promote the protection of investor legal rights ensuring equal treatment for domestic and foreign investors;

(f) Stimulate direct investments into capital intensive hydrocarbon projects by promoting profitable and efficient investments and adhering fully to the terms of international contracts and agreements;

(g) Establish joint ventures for the local manufacture and deployment of oil and gas equipment;

(h) Work collectively to enhance existing energy transport links, evaluate new energy transport corridors, establish energy transit agreements and develop environmentally sound energy transportation systems;

(i) Engage other UNECE member States in a dialogue on the ways and means of diversifying hydrocarbon exporting economies away from too much reliance on hydrocarbon resource exploitation and assess the investments needed to accomplish this in the context of sustainable development; and

(j) Work with other UNECE member States to improve the political, economic and social conditions within the region and in neighbouring countries by strengthening international relations and economic cooperation, enhancing economic development opportunities and addressing the causes of social and ethnic tensions.

2.2 COUNTRY PROFILES

This section was prepared on the basis of presentations by high-level officials and of national experts appointed by governments of the littoral Caspian Sea countries and of Turkey. These presentations were delivered to the High-Level Meeting on Energy Security in the Caspian Sea Region held in June 2005[48] (see also Annex) and at the Seminar on Energy Security Risk Mitigation and the Caspian Sea Region held in April 2006 under the auspices of the UNECE Committee on Sustainable Energy. This information was supplemented with data from the BP Statistical Review of World Energy June 2007; the United States Department of Energy, Energy Information Administration data base; and other sources.

This section complements Section 2.1 and provides additional information on the oil and gas industry and the hydrocarbon markets of the Republic of Azerbaijan, Islamic Republic of Iran, the Republic of Kazakhstan, the Russian Federation and Turkey.

<u>Azerbaijan</u>

Azerbaijan has a population of approximately 8.5 million. The country's real GDP grew by an impressive 25 per cent to about US$ 13 billion in 2005, as a result of higher oil and natural gas prices, an increase in production of hydrocarbons and continuing investment in major oil and gas projects. While the fuel sector represented 27 per cent of Azerbaijan's GDP in 2000, it accounted for 41 per cent of GDP in 2005.

According to the BP Statistical Review of World Energy, June 2007, Azerbaijan has proved oil reserves of one billion tonnes accounting for 0.6 per cent of the world's total and a reserves-to-production (R/P) ratio of 29.3. Oil production has steadily increased over the last ten years, from 9.2 Mt in 1995 to 14.0 Mt in 2000 and reaching a level of 32.5 Mt in 2006. Internal oil consumption over the same period contracted from 6.6 Mt in 1995 to 6.3 Mt in 2000 and 4.71 Mt in 2006. With increasing production and lower domestic demand, exports grew significantly over the period.

Proved reserves of natural gas are estimated at a level of 1.35 tcm, which corresponds to 0.7 per cent of the world total. Production has remained in the range of 5 - 6 bcm of gas per year since 1995, with 2006 production of 6.3 bcm. Over the period, domestic consumption first fell from 8.0 bcm in 1995 to 5.4 bcm in 2000 but has since climbed back to 9.4 bcm in 2006. Thus, despite significant reserves, Azerbaijan is currently a net natural gas importer and has a long-term supply contract with Gazprom.

Table 2.2.1 Estimated Oil Reserves and Production for Selected Countries

COUNTRY	RESERVES billion barrels End 2006	PRODUCTION barrels per day		
		2004	2005	2006
AZERBAIJAN	7.0	315 000	452 000	654 000
KAZAKHSTAN	39.8	1 297 000	1 356 000	1 426 000
TURKMENISTAN	0.5	193 000	192 000	163 000
TOTAL	47.3	1 805 000	2 000 000	2 243 000
RUSSIAN FEDERATION	79.5	9 287 000	9 552 000	9 769 000
IRAN	137.5	4 248 000	4 268 000	4 343 000

Source: BP Statistical Review of World Energy 2007.

There are currently four major oil and gas projects in the Republic of Azerbaijan that are of significance not only to the country's economy, but also for the European and world energy markets. These projects are the following:

- Full-scale development of the Azeri-Chyrag-Guneshli (ACG) oil field;
- Further development of the Shah-Deniz gas condensate field;
- Construction of the South Caucasus gas pipeline project Baku-Tbilisi-Erzurum (BTE); and
- The recently commissioned Baku-Tbilisi-Ceyhan (BTC) crude oil export pipeline.

By 2010 to 2012 these projects will contribute to significantly raising Azerbaijan's annual oil and natural gas production (according to some estimates, to 60 Mt of oil and 20 bcm of natural gas per year). National experts estimate that oil reserves of the Azeri-Chyrag-Guneshli (ACG) field alone are about one billion tonnes, plus 120 bcm of natural gas and 8 Mt of gas-condensate. The total investment for the development of the ACG field is expected to be about US$ 10 to 12 billion. Some 40 Mt of crude oil and 8 bcm of natural gas have already been produced from this field since it was commissioned in 1997.

National experts estimate proved reserves at the Shah-Deniz gas condensate field to be in the order of one tcm of gas and over 150 Mt of condensate. About US$ 4.0 billion will be invested during the first stage of development of this field. A double-function offshore platform for drilling and production will be installed at a sea level of 100 metres.

Since other offshore areas of the Caspian Sea, off Azerbaijan, are also very promising for new oil and gas field discoveries, the construction of an efficient and reliable pipeline network to bring Caspian oil and gas to the European and world markets has become a priority for Azerbaijan.

The first delivery of oil from Azerbaijan to world markets took place in October 1997 when the 1,400 km long Baku-Novorossiysk Pipeline was commissioned with a yearly delivery capacity of 6 Mt. The next step was the construction of the 850 km Baku-Supsa Pipeline, commissioned in April 1999, with a design capacity of over 6 Mt per year. In 1999, the Presidents of Azerbaijan, Georgia, Turkey and Kazakhstan signed the "Istanbul Declaration", which defined the major parameters of the Baku-Tbilisi-Ceyhan (BTC) Project aimed at the construction of an international pipeline, with a total length of 1,768 km and a design capacity of 50 Mt per year, that would deliver Caspian crude oil to the Mediterranean Sea. It was also envisaged that some crude oil from

Kazakhstan would be shipped through this pipeline. The BTC Project has now been commissioned with delivery of oil to Ceyhan.

Figure 2.2.1 The Routes of Existing and Planned Oil Pipelines

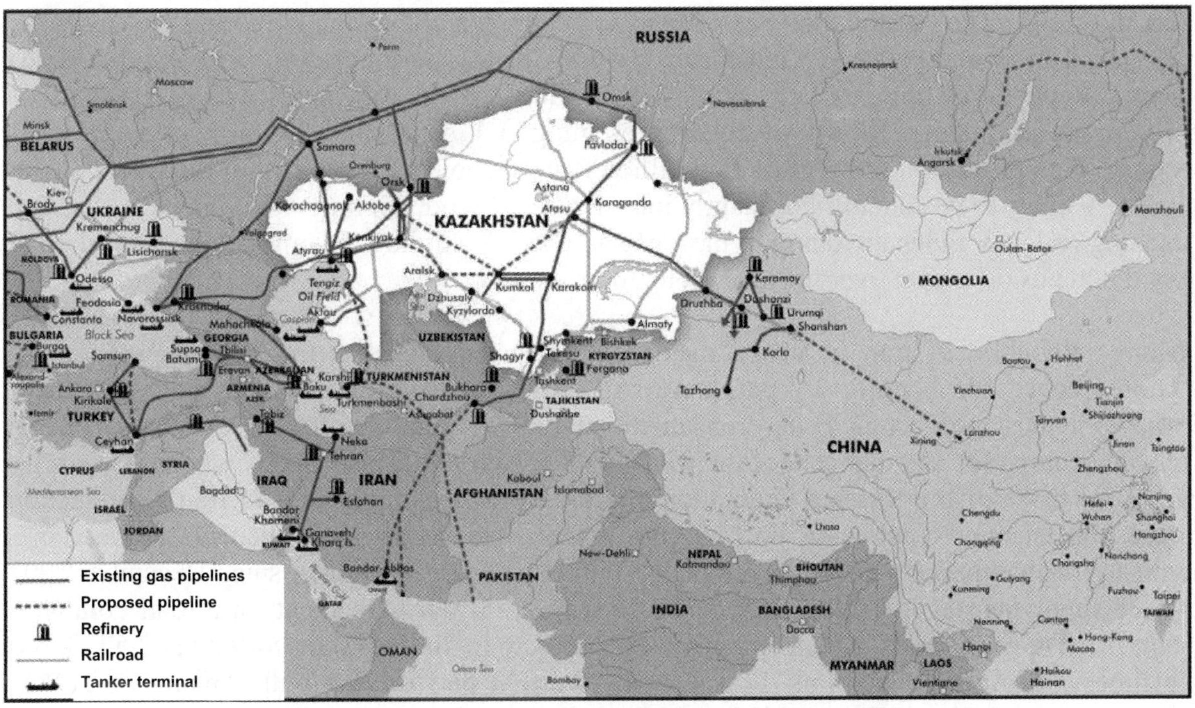

Source: International Energy Agency, Facts, Figures & Forecasts 2007, Caspian and Central Asian Perspectives

The South Caucasus gas pipeline connecting Baku, Tbilisi and Erzurum over a distance of some 915 km will bring gas produced at the Shah-Deniz field to Turkey and eventually to European gas markets. Pipeline construction began in late 2004 and is to be completed in 2007. BP and Norway's Statoil each have a 25.5 per cent stake in the project. The State Oil Company of Azerbaijan, SOCAR, Russia's Lukoil, the national oil and natural gas company of Turkey (TPAO), Total of France, and NICO of the Islamic Republic of Iran have about 10 per cent each. The investment costs are estimated at about US$ 1.3 billion and the pipeline is expected to carry over 20 bcm of natural gas per year.

To date, the Government of Azerbaijan has signed 25 oil agreements with 35 companies from 15 countries with a total investment estimated at over US$ 70 billion. About US$ 10 billion have already been invested in oil and gas infrastructure. In order to efficiently manage the revenues from oil and gas exports, in 1999 the Government of Azerbaijan established the State Oil Fund, which was designed to channel funds collected from oil and gas-related activities for education, poverty reduction and for efforts aimed at raising rural living standards. By the end of 2005, the State Oil Fund reported assets of US$ 1.3 billion and these are expected to have almost doubled during 2006.

Since 1991, the Government of Azerbaijan has been steadily implementing a modern and transparent legal and regulatory framework to promote the development of hydrocarbon projects, enhance foreign direct investment and generally promote economic development. Azerbaijan maintains close cooperation with various international organizations in the field of energy, including the UNECE, Energy Charter, the Organization of the Black Sea Economic Cooperation (BSEC) as well as being involved in special programmes of the European Union.

The Islamic Republic of Iran

The Islamic Republic of Iran, bordering the Caspian Sea as well as the Persian Gulf in the Middle East, is the only OPEC member among the Caspian Sea region countries and the second largest oil producer in OPEC. The country also has a strategic geographical location bordering the Caucasus, Central Asia, Turkey and Pakistan/Afghanistan (providing an export corridor to India and beyond). The Islamic Republic of Iran is a large crude oil producer with significant proven oil reserves. It currently produces about 210 Mt of oil, representing 5.4 per cent of total world crude oil production. With proved oil reserves of 18.9 billion tonnes, about 11.5 per cent of the world's total, the country is likely to be able to significantly increase its oil production in the future as it invests in both existing and new fields including in the Caspian Sea. According to some projections, it could possibly double its oil output over the next 15 to 30 years.

The Islamic Republic of Iran is also a major natural gas producer. In 2006, its natural gas production reached 105 bcm, or 3.7 per cent of the world total. However, it is not yet a major exporter, with most production being consumed internally. The country's proved reserves are 28.1 tcm accounting for 15.5 per cent of the world's total reserves. These are the second largest gas reserves after those of the Russian Federation and are very large compared to Iran's production. Therefore, the Islamic Republic of Iran has the potential to significantly increase production if economically viable markets can be developed. With regard to gas exports, since December 2001, it has been supplying gas to Turkey in accordance with a 25-year agreement for a total volume of 228 bcm. Iranian gas exports, which amounted to 5.69 bcm in 2006, is transported via 2,577 km of gas pipelines. These pipelines link Tabriz in the western part of Iran with Ankara in Turkey. Additionally, some 5.8 bcm of gas are imported from Turkmenistan.

While Iran is an important oil exporter, it is also a significant energy consumer. Domestic oil consumption has been growing steadily from 60 Mt in 1995 to 63.5 Mt in 2000 and 79.3 Mt in 2006. Domestic natural gas demand has been growing even more rapidly and in fact has almost tripled over the last ten years from 35.2 bcm in 1995 to 105 bcm in 2006. Currently, natural gas accounts for slightly more than half of Iran's total energy consumption. The Government is planning to make significant investments in the coming years to increase this share even further. The price of natural gas to residential and industrial consumers is state-controlled and kept at low levels. This has encouraged the rapid growth in natural gas consumption as a replacement for fuel oil, kerosene and liquefied petroleum gas (LPG).

The Islamic Republic of Iran is interested in fostering energy trade and cooperation among Caspian Sea countries. As a large regional energy consumer and with much of its hydrocarbon reserves located in the southern part of the country in and along the Persian Gulf, it is envisioning to import up to 10 bcm of gas from Turkmenistan by 2010 and up to 20 bcm by 2015, as well as 10 bcm from Azerbaijan provided that adequate energy transport infrastructure is available. Iran is also well placed to import Kashagan oil from Kazakhstan.

Kazakhstan

The Republic of Kazakhstan is both Central Asia's largest economy and one of the twenty largest oil-producing countries. Over the past ten years, oil production in the country has tripled. In 2006, oil and condensate production reached a level of 66 Mt. Given the large proved oil reserves of 5.5 billion tonnes, government authorities expect that oil production could reach the level of 90 Mt in 2010 and over 150 Mt by the year 2015.

Proved reserves of natural gas in Kazakhstan are 3.0 tcm, which represents 1.7 per cent of the world total. Kazakhstan produced 24 bcm of natural gas in 2006 and it plans to significantly raise production over the medium term. Government authorities expect that natural gas production could reach 52 bcm by 2010 and about 80 bcm by 2015.

Kazakhstan can accommodate rapid growth of hydrocarbon exports since it has abundant oil and gas reserves, relatively low domestic consumption of hydrocarbons compared to the reserve base (consumption of 11 Mt of oil and 20 bcm of gas in 2006) and domestic demand that appears to be rising only slowly. With this in mind and the growing global demand for hydrocarbon fuels, Kazakhstan's energy policy has been oriented towards the development of oil and gas projects, as well as the development of reliable oil and gas transportation infrastructure that can accommodate rising exports from Kazakhstan as well as from the other countries of the region to Western European and world markets.

KazMunayGas, the state oil and gas company of Kazakhstan, is currently involved in a number of ongoing and prospective export transportation projects aimed at opening up new markets, expanding existing export capacity and routes and diversify transportation networks. These projects include:

- Extension of the Caspian Pipeline Consortium (CPC) Project;
- Expansion of the capacity of the Atyrau–Samara oil pipeline;
- Linking Aktau in Kazakhstan with the Baku-Tbilisi-Ceyhan oil pipeline;
- Construction of the Atasu-Alashankou oil pipeline (as a precursor to the Kazakhstan-China pipeline project);
- Extension of "Central Asia–Centre" project;
- Feasibility study of the Aktau–Teheran (Kazakhstan-Turkmenistan-Iran) oil pipeline;
- Feasibility study of the gas pipeline "Kazakhstan–China"; and
- Assessment of participation in the Odessa–Brody–Plotsk oil pipeline system.

Figure 2.2.2 Kazakhstan Oil and Gas Facts

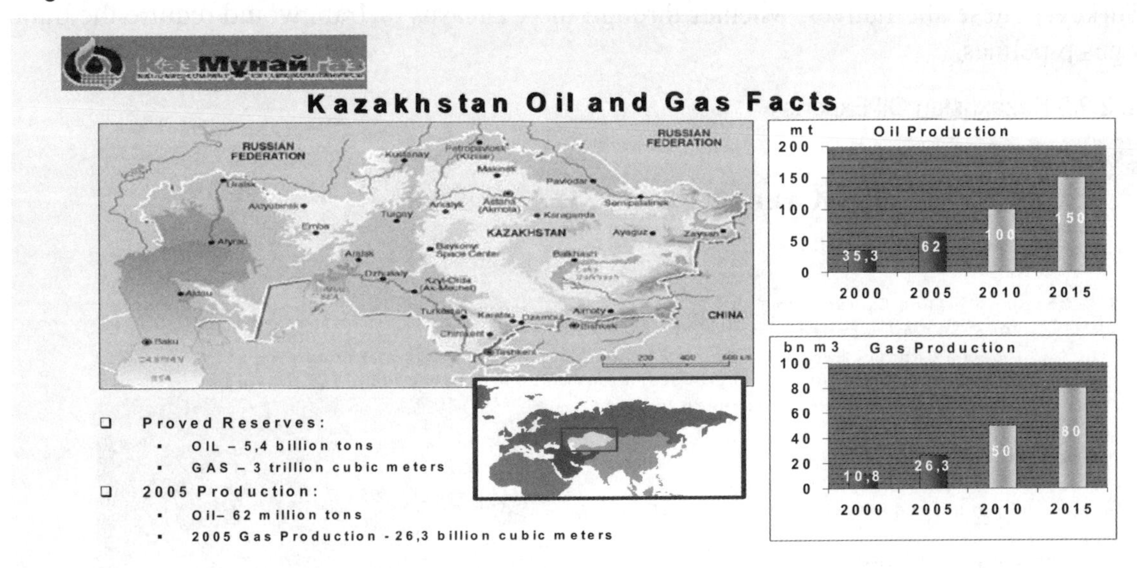

Source: Presentation by Kazakhstan at UNECE Seminar on Energy Risk Mitigation and the Caspian Sea Region, United Nations, Geneva 2006

Among the many projects listed above, the priority for Kazakhstan is to upgrade the capacity of the CPC pipeline system from 30.5 Mt in 2005 to 67 Mt of oil per year. This is a major export route for the country today. Since its commissioning in 2002, Kazakhstan has supplied over 80 Mt of oil through this pipeline. The second largest operating oil pipeline is the Atyrau–Samara pipeline from

Western Kazakhstan to Russia. It is foreseen that the throughput capacity of this pipeline will be expanded from 15 to 20-25 Mt per year.

Kazakhstan and Azerbaijan are currently considering the modalities of building a pipeline link from Aktau in Kazakhstan, across the Caspian Sea, to Baku in Azerbaijan to join up with the recently constructed Baku -Tbilisi-Ceyhan (ABTC) pipeline system, which would bring oil via the Caspian Sea to the BTC pipeline. At the end of 2005, the new Atasu-Alashankou oil pipeline was commissioned for supplying oil from Kazakhstan to China at a level of 10 Mt per year, with the expectation that the pipeline's throughput capacity would be expanded to 20 Mt per year at some later stage. KazMunayGas is also considering joining the Odessa-Brody-Plotsk Project to start deliveries of oil to Eastern and Central Europe and possibly even further to the Baltic.

The development of an efficient and reliable gas transportation system is a key component of Kazakhstan's energy policy. As noted earlier, gas production in Kazakhstan is steadily growing and could reach 80 bcm in 2015. Similarly, export volumes of gas are increasing from 7.6 bcm in 2005 to a projected level of 8.3 bcm in 2010.

Based on projected production and consumption of hydrocarbons and bearing in mind its geographical position, Kazakhstan has three major export corridors: the Western, Eastern and Southern corridors. At this time, Kazakhstan is paying particular attention to the Western corridor to serve the European gas market. Kazakhstan has two alternatives for serving this market. The first is an export corridor oriented in a northwesterly direction, using the existing high-pressure pipelines through the Russian Federation. In this case, the Soyuz and Oreburg–Novopskov pipelines would have to be uploaded with additional volumes of gas from the Karachaganah gas field and the capacity of the "Central Asia–Centre" gas transmission line would have to be upgraded from 54.8 bcm to 100 bcm per year. Alternatively, the Western corridor to serve the European gas market could be oriented in a southwesterly direction passing through the Caspian Sea and then onto the Caucasus and Turkey or directed in a southern direction through Iran and then on to Turkey. These alternatives, whether through the Caucasus or Iran, would require the building of new gas pipelines.

Figure 2.2.3 Kazakhstan Oil Exports

Source: Presentation by Kazakhstan at UNECE Seminar on Energy Risk Mitigation and the Caspian Sea Region, United Nations, Geneva 2006

Figure 2.2.4 The Key Transportation Vectors

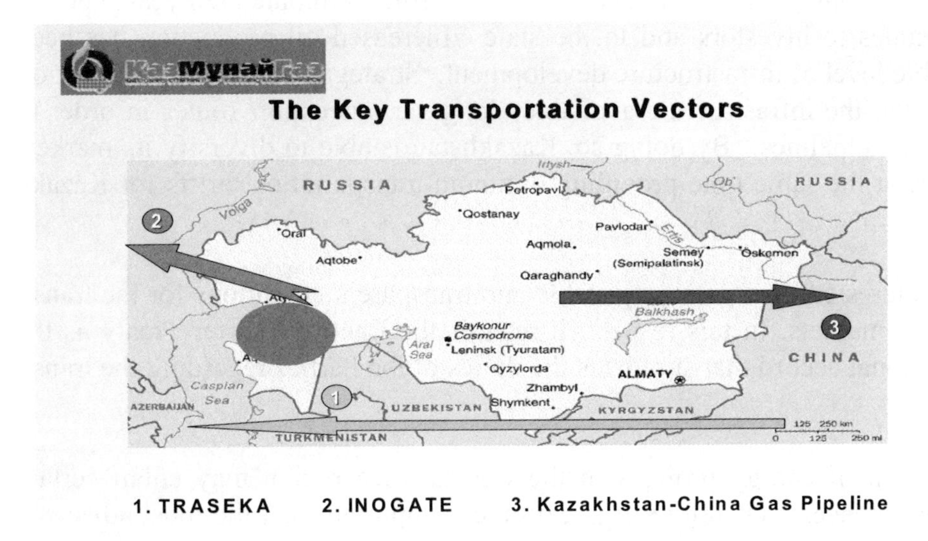

Source: Presentation by Kazakhstan at UNECE Seminar on Energy Risk Mitigation and the Caspian Sea Region, United Nations, Geneva 2006

The development of the south westerly gas transportation network is supported by the European Union Programme, Transport Corridor Europe-Caucasus-Asia (TRACECA). Similarly, the European Union member states are sponsoring the INOGATE programme, which supports development of oil and gas export corridors from Central Asian and the Caspian states to Europe (through Russia, Turkey or both).

The rapid economic growth and increased demand for energy in China has brought new opportunities to Kazakhstan in opening up new markets for its hydrocarbon exports. Therefore, KazMunayGaz is seriously considering the construction of an Eastern corridor. For their part, Chinese partners have already built a West–East gas pipeline as part of this future pipeline network.

Figure 2.2.5 Kazakhstan-China Pipeline

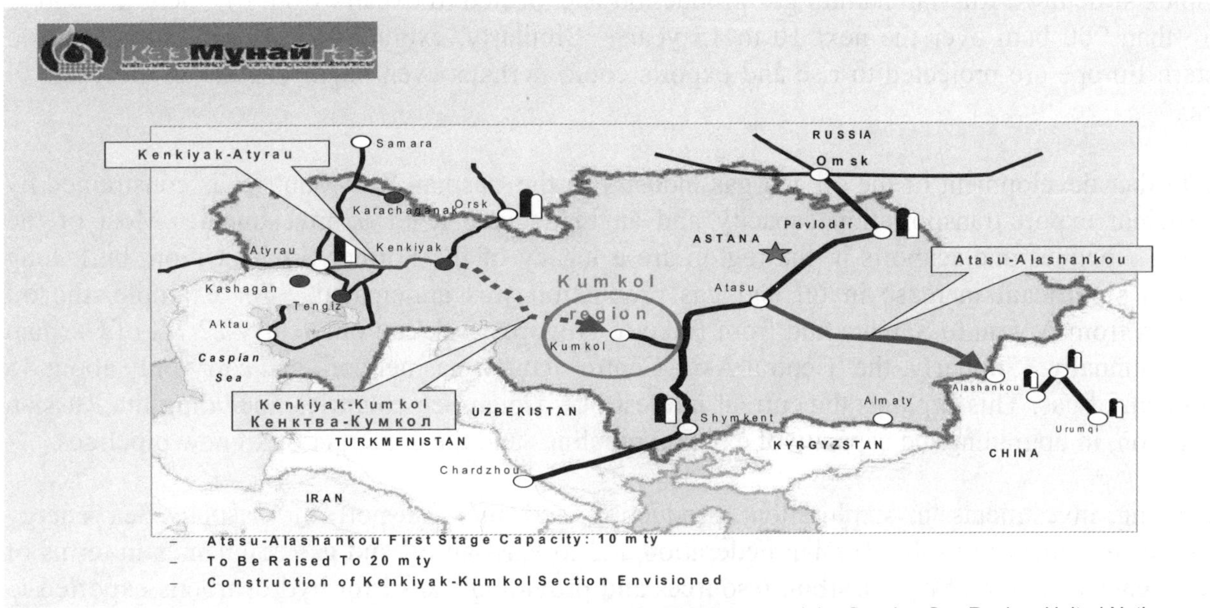

Source: Presentation by Kazakhstan at UNECE Seminar on Energy Risk Mitigation and the Caspian Sea Region, United Nations, Geneva 2006

Kazakhstan is successfully developing its oil and gas industries. Projects initiated ten years ago are being repaid, bringing revenues to investors and to the state. Increased oil production has been complemented with a suitable level of infrastructure development. Strategic decisions for Kazak oil exports take into consideration the infrastructure available along various export routes in order to optimize load volumes on key pipelines. By doing so, Kazakhstan is able to diversify its markets for oil and gas supply while at the same time providing optimum transportation tariffs for Kazakh oil and gas exporters.

Kazakhstan is particularly interested in maintaining stable and transparent conditions for the transit of energy resources to world markets. In this respect, it regards the Energy Charter Treaty as the only legally binding international accord that stipulates the duties of the Parties regarding the transit of energy resources.

The successful implementation of energy projects in the Caspian Sea region may entail certain adverse environmental consequences. Therefore, it is extremely important that the most advanced energy and resource-efficient technologies are applied in order to allow for the sustainable development of the region. This, in turn, calls for increased investment flows of advanced technologies to the countries of the Caspian Sea region.

Russian Federation

Over the last few years, the Russian Federation has significantly increased oil production, rising from 311 Mt in 1995, to 323 Mt in 2000 and 481 Mt in 2006. The Russian Federation accounts for around 12 per cent of total world oil production. Over the medium-term, production is expected to rise further to approximately 550 Mt per year. While much of this production is expected to come from existing fields, it is also anticipated that new greenfield projects will be developed in the Caspian Sea, East Siberia and Sakhalin.

The Russian Federation is the largest producer of natural gas in the world, accounting for about 21 per cent of total world gas production. It is also the country with the largest proved gas reserves, 47.7 tcm or 26 per cent of the world total. While new incremental gas supplies are likely to come from smaller fields, from more remote and harsher environments and from geologically more complex structures, Russian natural gas production is expected to expand from 612 bcm in 2006 to more than 700 bcm over the next 10 to 15 years. Similarly, exports to Western Europe and to Eastern Europe are projected to rise and exports could perhaps even begin to Asia in the years to come.

The further development of the oil and gas industry in the Caspian Sea countries is constrained by insufficient export transportation capacity and an inadequate level of investment. Most of the existing pipelines and seaports in the region are a legacy of the Former Soviet Union, built long before a significant increase in oil and gas production was anticipated. For example, the oil pipelines from Ätyrau to Samara and from Baku to Novorossiysk can only carry 22 Mt of Caspian Sea oil annually. Similarly, the 'Central Asia–Centre' natural gas network can carry only about 45 to 55 bcm of gas. This explains the current interest of Caspian Sea countries, including the Russian Federation, in upgrading the capacity of existing pipelines and the construction of new pipelines.

Enhancing investments in exploration, production and the transport of Caspian Sea energy resources is important to the Russian Federation and to Russian oil and gas companies in terms of investments, a source of hydrocarbon resources and providing transit for hydrocarbons exported to third countries. Leading Russian energy and oil companies operate jointly with other national energy companies in oil and gas projects in the Caspian Sea region. These joint ventures include the development of the Karachaganak and Kumkol oil fields in Kazakhstan, the Shah–Deniz gas condensate field in Azerbaijan, oil and gas fields in Uzbekistan, as well as a range of transportation projects.

The Russian oil company Lukoil began exploration of the Northern Caspian Sea in 1995. Five large oil and condensate fields have been found in this part of the Caspian Sea since Lukoil began exploring. These include Khvalinskoye, Yuri Korchagin, Rakushechnoye, and Sarmatskoye. The Khvalinskoye field will be developed by a fifty-fifty joint venture between Lukoil and Kazakhstan. In July 2003, Lukoil and Gazprom established a joint venture with Kazakhstan's state oil company, KazMunayGaz, to develop the Tsentralnaya hydrocarbon structure, located on the border of the Russian and Kazakhstani offshore sectors. The Tsentralnaya structure is estimated to hold recoverable reserves of approximately 0.6 tcm of natural gas. Gazprom is also a partner with KazMunayGaz in another project in the offshore Caspian Sea, the Kurmangazy project. The Kurmangazy field, which is estimated to contain around 1 billion tonnes of oil, is also located on the border between the Russian and Kazakh sectors of the Caspian Sea. Exploratory drilling at Kurmangazy began in 2003 with a total anticipated capital investment of US$ 2.1 billion.

Turkey

Energy demand in Turkey has been growing for decades at an annual rate of about 6 per cent, keeping pace with economic development and rising living standards. Today, domestic resources provide approximately 30 per cent of total energy demand while the remainder comes from a diversified range of oil, coal and natural gas imports.

Turkey has emerged as an important country in energy diplomacy and supply security, given its geographic location between western energy consuming markets and the large energy producers located in the Middle East and Caspian Sea region. Three quarters of the world's proved oil and gas reserves are located in these two regions adjacent to Turkey. One of the main elements of Turkey's energy strategy is to establish an energy corridor between the energy-rich countries of these two regions and the energy consuming markets, in particular Europe and the United States.

By providing an outlet as an energy transit country and a reliable distribution hub for hydrocarbon resources in the Caspian Sea region, Turkey contributes to:

(i) strengthen the independence and prosperity of the new Caspian Sea region states by ensuring the free flow of hydrocarbons to world markets;

(ii) encourage market economy and democratic developments;

(iii) stabilize the region by building up economic links among countries; and

(iv) diversify and secure energy supplies to Turkey and other consumers.

Figure 2.2.6 Transportation of Oil and Gas to Western Markets

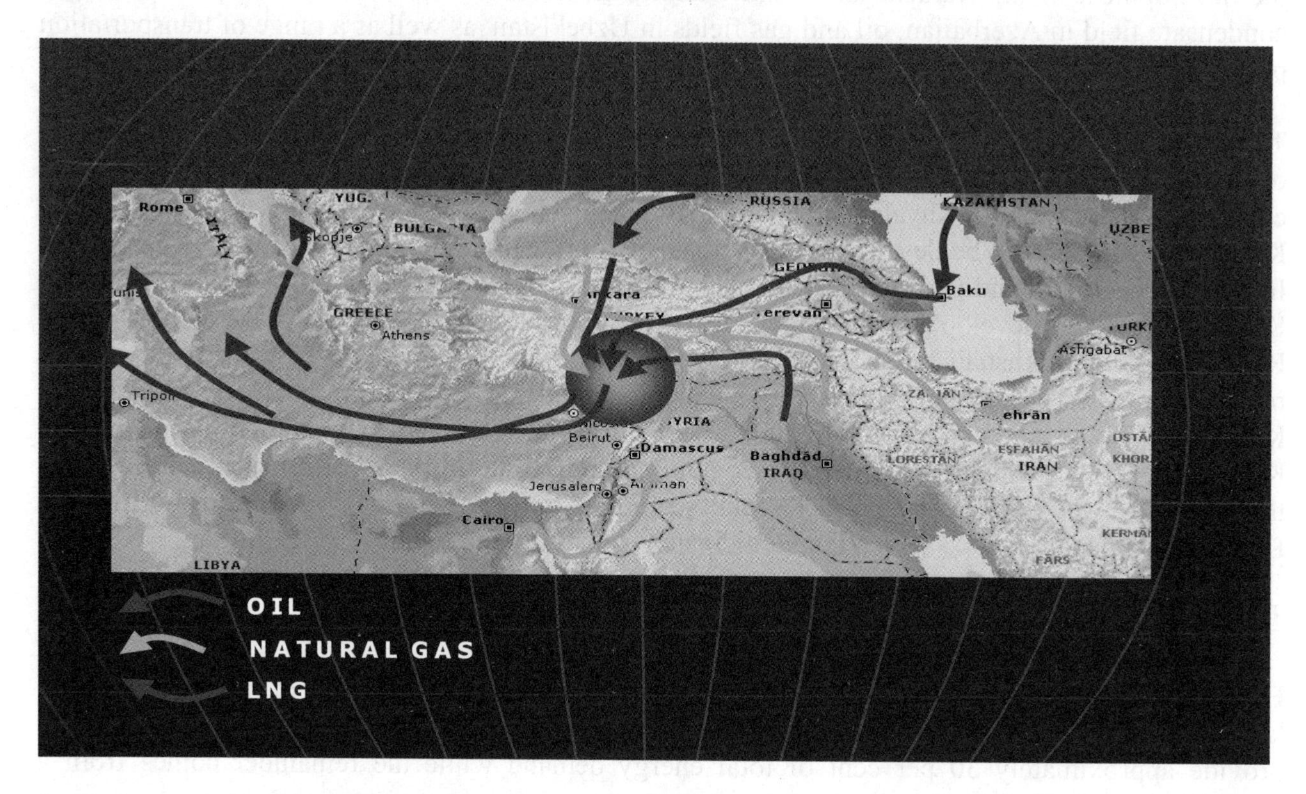

Source: Presentation by Republic of Turkey at UNECE Seminar on Energy Risk Mitigation and the Caspian Sea Region, United Nations, Geneva 2006

Within the concept of an East–West Energy Corridor, the Republic of Turkey participates actively in the implementation of two major projects that strengthen both regional and global energy security: the BTC oil pipeline project and the BTE natural gas pipeline project. Turkey has also joined initiatives to establish regional markets, such as the Energy Community of South-East Europe and the Mediterranean Ring Project. These initiatives are expected to increase cross-border electricity and gas trade and thus enhance energy security in the region.

Figure 2.2.7 Baku-Tbilisi-Ceyhan Main Export Pipeline Project

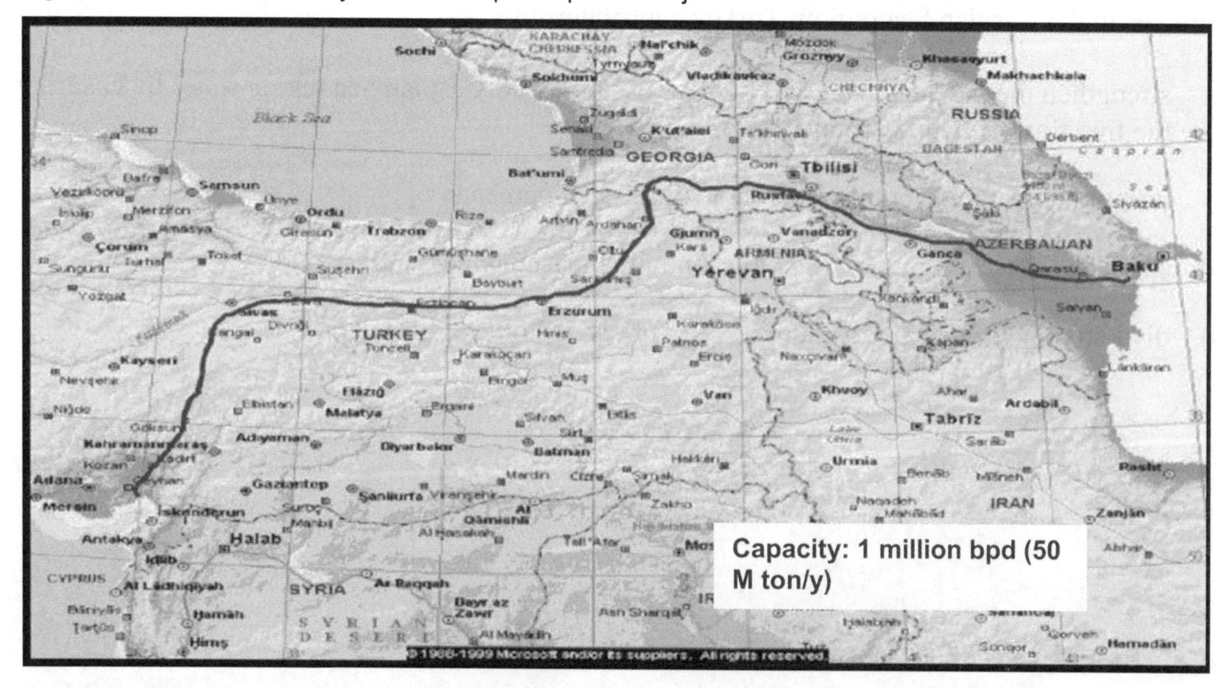

Source: Presentation by Republic of Turkey at UNECE Seminar on Energy Risk Mitigation and the Caspian Sea Region, United Nations, Geneva 2006

While Turkey is involved in developing projects to meet its own domestic demand, it is also involved in projects to meet the increasing energy demand of others, such as the European Union, by opening arteries to the European continent and to the Mediterranean Sea via Turkey. Three major new projects are currently under consideration: the Arab Natural Gas Pipeline to bring Middle East natural gas to western markets, the Turkey-Bulgaria-Romania-Hungary-Austria (NABUCCO) natural gas pipeline and the Turkish-Greek-Italian interconnection. In addition, the anticipated expansion of the BTC pipeline to Kazakhstan will consolidate Turkey's role as a hub and a transit country.

Some 6 to 7 per cent of global oil supply is expected to transit through Turkey by 2012 considering the capacities of the BTC pipeline, the Iraqi-Turkish oil pipeline from Kirkuk to Yumurtalik, the Samsun-Ceyhan oil by-pass pipeline now under development and the oil transported through the Bosphorus and Dardanelles straits. Furthermore, Turkey is likely to also become a major transit country and the fourth artery of the European Union for natural gas in the years to come. .

Figure 2.2.8 Nabucco Project

Source: Presentation by Republic of Turkey at UNECE Seminar on Energy Risk Mitigation and the Caspian Sea Region, United Nations, Geneva 2006

The European Union is the world's second largest market for natural gas, with demand projected to continue to increase. Therefore, it is understandable that the EU is seeking to diversify its sources of natural gas supplies although it already has a variety of suppliers – notably the Russian Federation, Norway and Algeria. In this context, natural gas from the Caspian Sea region offers an important prospect for meeting Europe's steadily increasing gas demand provided that secure transportation systems are established. Turkey offers such a secure route.

ANNEX

STATEMENT OF THE REPRESENTATIVES OF AZERBAIJAN, THE ISLAMIC REPUBLIC OF IRAN, KAZAKHSTAN, THE RUSSIAN FEDERATION AND TURKEY

High-Level Meeting on Energy Security in the Caspian Sea Region

Palais des Nations, Geneva
28 June 2005

During the UNECE Energy Security Forum High-Level Meeting on the Caspian Sea Region held on 28 June 2005, government representatives discussed emerging energy security risks and risk mitigation of the Caspian Sea region in a global context. In our statements at the meeting, a number of problems, challenges and opportunities were discussed:

1. The energy import dependence of most UNECE member countries will continue to rise in the foreseeable future, particularly for oil and natural gas, increasing their vulnerability to emerging energy security risks.

2. Western European oil imports could rise from 55 per cent of consumption now to 65 per cent in 2010 and possibly to 80 per cent by 2020. North American oil import dependence could rise from 35 per cent to 45 per cent by 2020. Apart from the Russian Federation, Central and Eastern European oil imports could rise from 80 per cent today to 90 per cent in 2020[1]. The oil import requirements of Asian countries, especially China and India, are also expected to increase significantly.

3. Western European natural gas imports are expected to rise from 35 per cent of consumption now to 45 per cent in 2010. United States liquefied natural gas (LNG) imports are likely to continue rising. Apart from the Russian Federation, Central and Eastern European gas import dependence is likely to increase from 65 per cent in 2010 to 85 per cent in 2015.

4. The traditional suppliers of natural gas from the region, such as the Islamic Republic of Iran and the Russian Federation, are likely to have the capacity to meet Europe's growing demand but significant new investments will be needed for production and transportation infrastructure.

5. While oil production in the Caspian Sea region of Azerbaijan, Kazakhstan, the Russian Federation and Turkmenistan is currently a small proportion of world oil production, there is a large potential for the future. Oil production is likely to increase in the Caspian Sea Region and even possibly double over the next five to ten years. Natural gas production and gas exports are also expected to rise in the coming years.

6. In order to accommodate increased exports from the Caspian Sea region, developed and enhanced transport facilities and new planned transport corridors will be further considered together with transit rights, new production and transportation systems. The large investments required to accomplish those projects also demand a suitable investment climate, commercial agreements and political will.

[1] World Energy Outlook, 2004, International Energy Agency, Paris.

7. Together with existing pipeline systems, the Baku-Tbilisi-Ceyhan Crude Oil Petroleum Pipeline that will be operational in the fourth quarter of 2005 and the South Caucasus natural gas pipeline project (Baku-Tbilisi-Erzurum natural gas pipeline project), which is expected to be realised in 2006, are the two significant elements of the East-West Energy Corridor as the new transport corridors which will enhance energy security. Likewise, alternative transport corridors from the Caspian Sea region via the Islamic Republic of Iran to international markets will also contribute to the diversity of energy transport routes and energy security.

After our dialogue, as representatives of the countries concerned, we acknowledge that:

1. The diversification of energy trading partners, international cooperation and producer-consumer dialogue are key policy options that UNECE member States can pursue together to mitigate emerging energy security risks.

2. The doubling of energy production in the Caspian Sea region during the next five to ten years and therefore greater exports could provide UNECE energy importing member States with additional energy supply options, thus contributing to mitigation of energy security risks across the UNECE regional energy market.

3. In order for UNECE member States to benefit from the increased oil and natural gas exports from the Caspian Sea region, multi-billion dollar investments will be needed to expand energy production capacities and provide for new energy transport infrastructure. Countries of the Caspian Sea region will need to establish and maintain a suitable investment environment, regulatory framework, facilitating the transfer of technology, as well as provide unrestricted access of their energy products to the European market.

4. In order to accommodate the potential increase in energy exports from the Caspian Sea region, the countries concerned will work together and with our partners on enhanced transport facilities, new transport corridors, transit agreements, new energy production and transportation systems while enhancing the environmental quality of the region in accordance with the principles of sovereignty of States over their natural resources and economic activity and freedom to choose a suitable framework for their foreign economic relations.

5. The benefits associated with increased oil and gas exports from the Caspian Sea region would contribute to our efforts to achieve sustainable development and fulfil the Millennium Development Goals.

Representatives of the countries concerned welcomed the proposal of the Energy Security Forum to work together with national experts appointed by the government of each participating country as well as individual experts to:

1. Prepare a study on emerging energy security risks and risk mitigation in a global context, including the potential contribution of increased energy exports of the Caspian Sea region to provide greater diversity of energy supply sources to UNECE member States;

2. Convene a seminar to examine energy transport corridors, new infrastructure, transmission systems and investment requirements for increased energy exports from the Caspian Sea region including conclusions and recommendations on enhanced energy trade and international cooperation; and

3. Submit the conclusions and recommendations of the study and seminar to a subsequent session of the Committee on Sustainable Energy for a decision on further transmission to the relevant bodies of the United Nations system and to recommend follow-up activities for international cooperation on energy security.

REFERENCES, BIBLIOGRAPHY AND COMMENTS

Section 1.3: Energy Security Risks and Risk Mitigation: A Global Overview

References

1. *International Energy Agency, World Energy Outlook 2006, Paris.*

Section 1.4 Energy Security and the European Union

References

2. *Various publications of the International Energy Agency, Paris.*
3. *Yergin D., "Energy Security and Markets," in Energy and Security: Toward a New Foreign Policy Strategy.*
4. *Cambridge Energy Research Associates, Energy Watch, Spring 2005.*
5. *Wardell S., Global Insight - personal communication.*
6. *Hueper P., Fundamentals of Energy Infrastructure Security, Petroleum Economist 2005.*
7. *International Energy Agency, World Energy Outlook 2006, Paris.*
8. *European Commission, Energy Green Paper, SEC (2006) 317/2, Brussels.*
9. *United Kingdom Cabinet Office, 'Countries at Risk of Instability' Process Manual, May 2005, London.*
10. *Alington N., AON, Westminster Energy Forum Conference, April 2006, London.*

Section 1.5: An Eastern Perspective of Energy Security

11. *Various publications of the International Energy Agency, including Key World Energy Statistics 2006, Paris.*
12. *BP Statistical Review of World Energy June 2007, London.*
13. *International Energy Agency, World Energy Outlook 2006, Paris.*
14. *International Energy Agency, Key World Energy Statistics 2006 and World Energy Outlook 2006, Paris.*
15. *International Energy Agency, World Energy Outlook 2005 and 2006, Paris.*
16. *European Commission, Energy Green Paper, SEC (2006) 317/2, Brussels.*
17. *Karpus P. and D. Bocharov, Strategy Cannot Be Bureaucratic, World Power, Number 11, November 2005.*
18. *Schlessinger J., Thinking Seriously Future, The National Interest, Number 82, Winter 2005/2006.*
19. *Kokoshin A. A., The Issues of Supporting the Global Energy Security, Institute of International Security, Russian Academy of Sciences, 2006, Moscow.*
20. *G8 Statement, "Global Economy and Oil", Gleneagles, 2005.*
21. *Kokoshin A. A., The Issues of the International Energy Security and the Policy of Russia, Institute of International Security, Russian Academy of Sciences, 2005, Moscow.*
22. *Zweig D., Jianhai B., China's Global Hunt for Energy, Foreign Affairs, Sept./Oct. 2005. Vol. 84. No. 5.*
23. *Nekrasov A. S., Economic Problems and Future Outlook of Russia's Energy Sector, Presentation at the general meeting of the Russian Academy of Sciences, 20 December 2005.*
24. *Presentation by A. Miller, Chairman of "Gazprom" Board, at the annual meeting of "Gazprom" PJSC shareholders, 24 June 2005.*
25. *European Commission, Green Paper, "A European Strategy for Sustainable, Competitive and Secure Energy", Brussels, 8 March 2006.*
26. *Presentation by I. Osokina, Deputy Minister of Economic Development and Trade of the Russian Federation, at the Conference "CERA-2004" in Houston, Texas, United States, 10 February 2004.*
27. *Zakharova T., Oil and Gas Production: Basis for Russia's Security, Neftegaz.*
28. *Presentation by V. Shkolnik, Minister of Energy and Mineral Resources of the Republic of Kazakhstan, in Geneva on June 28, 2005.*
29. *Kokoshin A. A., Energy Cooperation Between Russia and China Is Promising, RIA "Novosti," January 12, 2005.*
30. *Speech by V. V. Putin at the Conference of the Security Council of the Russian Federation, 22 December 2005.*
31. *Kokoshin A. A., Real Sovereignty in the Contemporary Political System of the World, KomKniga, 2005, Moscow.*

Section 1.6: The North American View of Energy Security

References

32. *International Energy Agency, World Energy Outlook 2006, Paris.*
33. *BP Statistical Review of World Energy June 2007, London.*
34. *United States Department of Energy, Energy Information Administration, Annual Energy Outlook 2006.*
35. *United States Department of Energy, Energy Information Administration, statistical data base.*
36. *ICF Consulting, "Refinery Testimony", October 19, 2005.*
37. *UNCTAD World Investment Report 2005.*
38. *United Nations Statistics, statistical data base.*
39. *International Energy Agency, Oil Fact Sheet 2005.*
40. *Petroleum Economist, January 2006.*
41. *RAND Corporation, Oil shale development in the United States, 2005.*
42. *International Energy Agency, Renewable Energy: Market and Policy Trends in IEA Countries, 2004, Paris.*
43. *International Energy Agency, Renewables for Power Generation: Status and Prospects, 2003, Paris.*
44. *United States Department of Energy, Energy Information Administration, "Canada Country Analysis Brief," February 2005.*
45. *LNG Journal, November/December 2005.*

Section 2.1 Report on Global Energy Security and the Caspian Sea Region

References

46. *Various sources, including International Energy Agency, World Energy Outlook 2006.*
47. *BP Statistical Review of World Energy June 2007, London.*

Section 2.2 Country Profiles

References

48. *United Nations eBook publication, Energy Security in the Caspian Sea Region, ECE Energy Series No. 35, 2006, New York and Geneva. Also see United Nations eBook publication, New Energy Security Threats, ECE Energy Series No. 19, 2003, New York and Geneva.*

****